T0235542

Microwave Integrated Circuit Components Design through MATLAB®

Microwave Integrated Circuit Components Design through MATLAB®

S. Raghavan

CRC Press
Taylor & Francis Group
Boca Raton London New York

CRC Press is an imprint of the
Taylor & Francis Group, an **informa** business

MATLAB® is a trademark of The MathWorks, Inc. and is used with permission. The MathWorks does not warrant the accuracy of the text or exercises in this book. This book's use or discussion of MATLAB® software or related products does not constitute endorsement or sponsorship by The MathWorks of a particular pedagogical approach or particular use of the MATLAB® software

CRC Press
Taylor & Francis Group
52 Vanderbilt Avenue,
New York, NY 10017

First issued in paperback 2021

© 2020 by Taylor & Francis Group, LLC
CRC Press is an imprint of Taylor & Francis Group, an Informa business

No claim to original U.S. Government works

ISBN-13: 978-0-367-24312-8 (hbk)
ISBN-13: 978-1-03-208499-2 (pbk)

This book contains information obtained from authentic and highly regarded sources. Reasonable efforts have been made to publish reliable data and information, but the author and publisher cannot assume responsibility for the validity of all materials or the consequences of their use. The authors and publishers have attempted to trace the copyright holders of all material reproduced in this publication and apologize to copyright holders if permission to publish in this form has not been obtained. If any copyright material has not been acknowledged please write and let us know so we may rectify in any future reprint.

Except as permitted under U.S. Copyright Law, no part of this book may be reprinted, reproduced, transmitted, or utilized in any form by any electronic, mechanical, or other means, now known or hereafter invented, including photocopying, microfilming, and recording, or in any information storage or retrieval system, without written permission from the publishers.

For permission to photocopy or use material electronically from this work, please access www.copyright.com (http://www.copyright.com/) or contact the Copyright Clearance Center, Inc. (CCC), 222 Rosewood Drive, Danvers, MA 01923, 978-750-8400. CCC is a not-for-profit organization that provides licenses and registration for a variety of users. For organizations that have been granted a photocopy license by the CCC, a separate system of payment has been arranged.

Trademark Notice: Product or corporate names may be trademarks or registered trademarks, and are used only for identification and explanation without intent to infringe.

Library of Congress Control Number: 2019949280

Visit the Taylor & Francis Web site at
http://www.taylorandfrancis.com

and the CRC Press Web site at
http://www.crcpress.com

Dedicated to

JEGADHA Singaravelus,
Alagappans, Sarojini Sellamuthus, VNC Vijayakumars
Dr. A. MEENAKSHI SUNDARI, ARUNDHATHI, ABHIMANYU
Students and
Prof. Bharathi Bhat

Contents

Foreword

The evolution of "microwave integrated circuits" (MIC) over the past five decades or so has revolutionized the very approach to microwave components/system design and technology. As a result, the earlier waveguide-based components/systems have been, almost completely, taken over by the lighter and more compact MIC-based versions. It is this area of MIC that S. Raghavan has been involved with throughout his professional career at the National Institute of Technology-Trichy (India).

Besides teaching at the BE/MTech levels, Dr. Raghavan has been guiding research and technology-driven projects in the area of MIC. Realizing the importance of practical knowledge, he has created advanced laboratories so that the students can have hands-on experience in the design, development, and testing of MIC components and antennas. He has also evolved MATLAB-based techniques to ease the design of various MIC components. This book is the outcome of his dedicated effort and involvement in teaching and research in this specialized area for nearly 35 years.

The book emphasizes the network approach to the design of MIC components, and this aspect has been lucidly explained. The application of this approach to the most commonly used components forming any MIC system such as, filters, directional couplers, power dividers, and amplifiers is covered systematically. In a way, the seemingly difficult subject has been simplified through the use of circuit theory and MATLAB-based design techniques. With its easy-to-understand style, this book should be useful not only to the students and designers of MIC but also to students having the basic knowledge of network theory to understand integrated circuits at microwave frequencies.

Bharathi Bhat
Retired Professor, IIT Delhi

Preface

Necessity is the mother of invention. This is the philosophy behind numerous inventions. During the Second World War, there was a dire necessity for a high-power oscillators. The solution came in the form of the magnetron. Due to the skin depth, the conducting wires were of no use at microwave frequencies and so came the usage of waveguides. As the frequency gets higher and higher, the waveguides also find lesser usage in fewer applications. Then, the need for microwave integrated circuits (MICs) was felt technically. Not only that, due to the inventions of smaller devices like Read, Avalanche Transit Time devices, and other similar diodes, the waveguides could not be used for integration. The backbone of MICs are planar transmission lines. First in that family is the strip line, followed by the micro strip line, slot line, coplanar waveguides (CPWs), coplanar strip (CPS), finline, image guide, and variants of the abovementioned planar transmission lines. The golden period of microwaves (centimeter waves) was 1935–1945. The transmission lines being distributed networks, the first operation one should do is to find out the equivalence of distributed theory into lumped theory. The first chapter deals with a thorough microwave circuit theory along with relevant examples. The equivalent network theory needed for the design is also explained. The design parameters and characteristics of various planar transmissions lines are detailed in the second chapter.

The third chapter contains the basics of even mode and odd mode analysis of the microwave integrated components. From the first principle, the design formulae derivation and layout formulation of every component (branch-line coupler, hybrid ring coupler, rat-race coupler, backward wave coupler, and various power dividers) are detailed. Filters are very important components. If a person is thorough with the design and layout of the filters, understanding amplifiers and oscillators can become very simple. Binomial theory, maximally flat response, filter theory, determination of prototype values, and design and layout of various filters form the fourth chapter. The fifth chapter contains stability analysis, various power gains, constant gain circles, stability circles, noise figure circles, and the design and layout of matching networks for the amplifiers. The same theory with the Rollet stability factor and the Delta factor having an opposite magnitude of that of the amplifiers will facilitate one to be thorough with microwave oscillator design procedures. All the above analysis is supported by conventional formulae (from the first principle).

The conventional basic formulae are used for the entire design of techniques in all chapters. If one understands the theory and knows the ins and outs of MATLAB®, the programming (CAD DESIGN) can very easily be done. THE BEAUTY and (for the first time) the UNIQUE PURPOSE OF

THIS BOOK is that every step explained in the theory portion is also written in MATLAB. Those programs are tested and one can run and verify them. This process will inspire one to write one's own program and make the microwave integrated circuit component design an interesting one. Microwaves are made easy by proper understanding of the concepts. They are made easy by easy understanding! Who said microwave designs are tough and unpopular? This book is the outcome of 38 years of teaching in REC/NIT, Trichy by the author.

Dr. S. Raghavan
NIT, Tiruchirappalli
30.4.19

MATLAB® is a registered trademark of The MathWorks, Inc. For product information, please contact:
The MathWorks, Inc.
3 Apple Hill Drive
Natick, MA 01760-2098 USA
Tel: 508-647-7000
Fax: 508-647-7001
E-mail: info@mathworks.com
Web: www.mathworks.com

Acknowledgments

Prof. S. Raghavan gratefully thanks Prof. Bharthi Bhat, Prof. S. K. Koul, and every author of the reference books.

He also wishes to record the logistics work carried out by Dr. Samson Daniel.

Author

S. Raghavan was a Senior Professor (H.A.G.) in the Electronics and Communication Engineering Department, National Institute of Technology (NIT), Trichy until July 2019, and has 38 years' teaching and research experience. His interests include: microwave integrated circuits, RF MEMS, BioMEMS, metamaterial, frequency-selective surfaces (FSS), and microwave engineering. He has established a state-of-the-art microwave integrated circuit and microwave laboratory at NIT, Trichy with the help of funding from the Indian government. He has won Best Teacher award twice and has been conferred with the Honorary Fellowship of Ancient Sciences and Archaeological Society of India. Prof. Raghaven was a visiting fellow at California State University, North Ridge, US. He has conducted a tutorial at APEMC 2010, Beijing, China and served as the Organizing Chair of the Indian Antenna Week 2014, Chandigarh. Prof. Raghaven was invited to be a session chair at the PIERS 2013 symposium, Taipei, Taiwan. He has 90 research papers in international journals, 80 in the IEEE Xplore digital library, 130 international conferences, and 26 national conferences to his credit. To date, he has guided more than 12 PhD scholars. Prof. Raghaven is a Senior Member/Fellow in more than 22 international and national professional societies including: IEEE, IEI, IETE, CSI, ISSS, and TSI. Apart from organizing various workshops of national importance, he has also greatly contributed to the development of a state-of-the-art library and hospital in NIT, Trichy. Prof. Raghaven was the founding chairman, student branch of IEEE (MTT) and IEEE (APS) of NIT, Trichy. He was a patron for the most successful international IEEE conference (INICPW 2019, May 2019), held for the first time in ECE's 56-year history at REC/NIT, Trichy. He is a Senior Member of IEEE (MTT, APS, and EMBS), and is a recipient of many awards conferred based on his contributions to microwave engineering. The Institution of Electronics and Telecommunication Engineers (IETE) (India), has awarded Prof. S. Raghavan 'Smt. Ranjana Pal Memorial Award 2019' for his contribution to; Microwave Integrated Circuits, Frequency Selective Surfaces (FSS), Substrate Integrated Waveguides (SIW), Microwave Integrated Circuit Design Essentials through MATLAB and Popularization of Microwaves among student community.

1

Transmission Line Networks

1.1 Introduction

A transmission line can be considered as a distributed two-port networks with V_S, V_R and I_S, I_R of distributed network as equivalent to V_1, V_2 and I_1, I_2 of a lumped two-port networks. By this, one can easily obtain ABCD parameters which are otherwise called transmission parameters and also known as chain parameters (Figure 1.1).

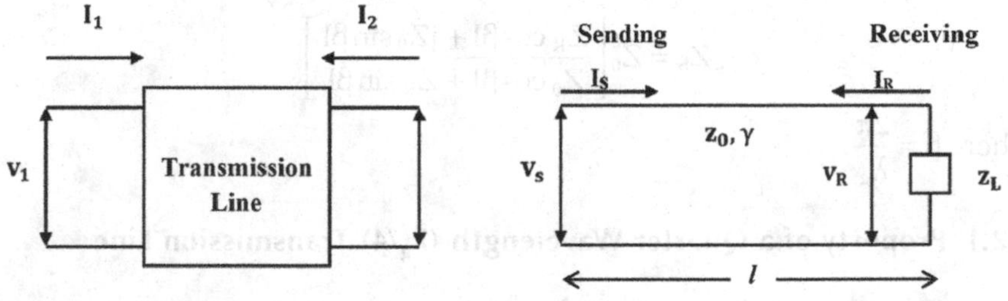

FIGURE 1.1
Two-port network representation of a transmission line.

$$\text{Sending end impendence } Z_S = \frac{V_S}{I_S}$$

$$\boxed{Z_S = Z_0\left[\frac{Z_R \cosh \gamma l + Z_0 \sinh \gamma l}{Z_0 \cosh \gamma l + Z_R \sinh \gamma l}\right]}, \tag{1.1}$$

where

Z_0 = Characteristic impedance = 50 Ω (Unless or otherwise specified)

$\gamma = \alpha + j\beta$

α = attenuation constant in Nepers

β = propagation constant in radians

l = length of the transmission line.

For a lossless transmission line, $\alpha = 0$. Therefore, $\gamma = j\beta$. Equation (1.1) then becomes

$$\boxed{Z_S = Z_0\left[\frac{Z_R\cos\beta l + jZ_0\sin\beta l}{Z_0\cos\beta l + Z_R\sin\beta l}\right]} \quad (1.2)$$

1.2 Characteristic Impedance for Different Lengths ($\lambda_g/4$, $\lambda_g/2$, $\lambda_g/8$)

For a lossless transmission line network, the characteristic impedance is given by

$$Z_S = Z_0\left[\frac{Z_R\cos\beta l + jZ_0\sin\beta l}{Z_0\cos\beta l + Z_R\sin\beta l}\right]$$

where $\beta = \dfrac{2\pi}{\lambda_g}$.

1.2.1 Property of a Quarter Wavelength ($\lambda_g/4$) Transmission Line

$$\text{If } l = \frac{\lambda_g}{4} \text{ then } \beta l = \frac{\pi}{2} \quad (1.3)$$

where

$\lambda_g = \dfrac{\lambda_0}{\varepsilon_{eff}}$

λ_g = guide wavelength

ε_{eff} = effective dielectric constant

Substituting Equation (1.3) in (1.2) we get

$$Z_S = Z_0\left[\frac{Z_R\cos\left(\frac{\pi}{2}\right) + jZ_0\sin\left(\frac{\pi}{2}\right)}{Z_0\cos\left(\frac{\pi}{2}\right) + jZ_R\sin\left(\frac{\pi}{2}\right)}\right]$$

$$\boxed{Z_S = \frac{Z_0^{\ 2}}{Z_R} \text{ (or) } Z_S Z_R = Z_0^{\ 2}} \tag{1.4}$$

The $\lambda_g/4$ transmission line is called an impedance transformer and also an impedance inverter because of Equation (1.4).

1.2.2 Property of a Half Wavelength ($\lambda_g/2$) Transmission Line

$$\text{If } l = \frac{\lambda_g}{2} \text{ then } \beta l = \pi. \tag{1.5}$$

Substituting Equation (1.5) in (1.2)

$$Z_S = Z_0 \left[\frac{Z_R \cos(\pi) + jZ_0 \sin(\pi)}{Z_R \sin(\pi) + jZ_0 \cos(\pi)} \right]$$

$$\boxed{Z_S = Z_R}. \tag{1.6}$$

The $\lambda_g/2$ transmission line is called an impedance repeater.

1.2.3 Property of a One-Eighth Wavelength ($\lambda_g/8$) Transmission Line

$$\text{If } l = \frac{\lambda_g}{8} \text{ then } \beta l = \frac{\pi}{4} \tag{1.7}$$

Substituting Equation (1.7) in (1.2)

$$Z_S = Z_0 \left[\frac{Z_R \cos\left(\frac{\pi}{4}\right) + jZ_0 \sin\left(\frac{\pi}{4}\right)}{Z_0 \cos\left(\frac{\pi}{4}\right) + jZ_R \sin\left(\frac{\pi}{4}\right)} \right]$$

$$\boxed{|Z_S| = |Z_0|} \tag{1.8}$$

1.3 T-Network and π-Network Sections Equivalent of a Transmission Line

See Figure 1.2.

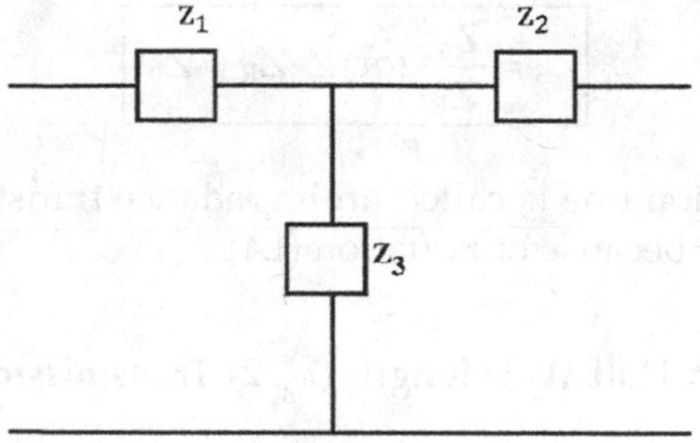

FIGURE 1.2
T-network.

$$Z_{1oc} = Z_1 + Z_3 \tag{1.9}$$

$$Z_{1sc} = Z_1 + \frac{Z_2 Z_3}{Z_2 + Z_3} \tag{1.10}$$

$$Z_{2oc} = Z_2 + Z_3 \tag{1.11}$$

$$Z_{2sc} = Z_2 + \frac{Z_1 Z_3}{Z_1 + Z_3} \tag{1.12}$$

Subtracting Equation (1.9) from (1.10)

$$Z_{1oc} - Z_{1sc} = Z_1 + Z_3 - Z_1 - \frac{Z_2 Z_3}{Z_2 + Z_3}$$

$$Z_{1oc} - Z_{1sc} = \frac{Z_3{}^2 + Z_2 Z_3 - Z_3}{Z_2 + Z_3} = \frac{Z_3{}^2}{Z_2 + Z_3}$$

$$Z_{1oc} - Z_{1sc} = \frac{Z_3{}^2}{Z_{2oc}} \left[\text{by } 1.21\right]$$

$$Z_3 = \sqrt{Z_{2oc}\left(Z_{1oc} - Z_{1sc}\right)}. \tag{1.13}$$

We know that $Z_{oc} = \dfrac{Z_0}{j\tan\beta l}$ and $Z_{sc} = jZ_0 \tan \beta l$.

Substituting Z_{oc} and Z_{sc} values in Equation (1.13)

$$Z_3 = \sqrt{\frac{Z_0}{j\tan\beta l}\left(\frac{Z_0}{j\tan\beta l} - jZ_0\tan\beta l\right)}$$

$$= \sqrt{\frac{Z_0}{j\tan\beta l}\left(\frac{Z_0 + Z_o\tan^2\beta l}{j\tan\beta l}\right)}$$

$$= \sqrt{\frac{{Z_0}^2}{j^2\tan^2\beta l}\left(1+\tan^2\beta l\right)}$$

$$= \sqrt{\frac{{Z_0}^2}{j^2\tan^2\beta l}\left(\sec^2\beta l\right)} = \frac{Z_0}{j\sin\beta l}.$$

From (1.90) and (1.91)

$$\boxed{Z_3 = \frac{Z_0}{j\sin\beta l}} \tag{1.14}$$

Substituting Equation (1.14) in (1.9)

$$Z_1 = Z_{1oc} - Z_3$$

$$= \frac{Z_0}{j\tan\beta l} - \frac{Z_0}{j\sin\beta l}$$

$$= \frac{Z_0\cos\beta l}{j\sin\beta l} - \frac{Z_0}{j\sin\beta l} = \frac{Z_0\left(\cos\beta l - 1\right)}{j\sin\beta l}$$

$$= -\frac{Z_0}{j}\frac{2\sin^2\frac{\beta l}{2}}{2\sin\frac{\beta l}{2}\cos\frac{\beta l}{2}} = -\frac{Z_0}{j}\tan\frac{\beta l}{2}$$

$$\boxed{Z_1 = jZ_0\tan\frac{\beta l}{2}} \tag{1.15}$$

Similarly, from Equation (1.11)

$$\boxed{Z_2 = jZ_0\tan\frac{\beta l}{2}} \tag{1.16}$$

Substituting Equations (1.14)–(1.16) in the T-network, we get Figure 1.3a. Similarly, for π-network, see Figure 1.3b.

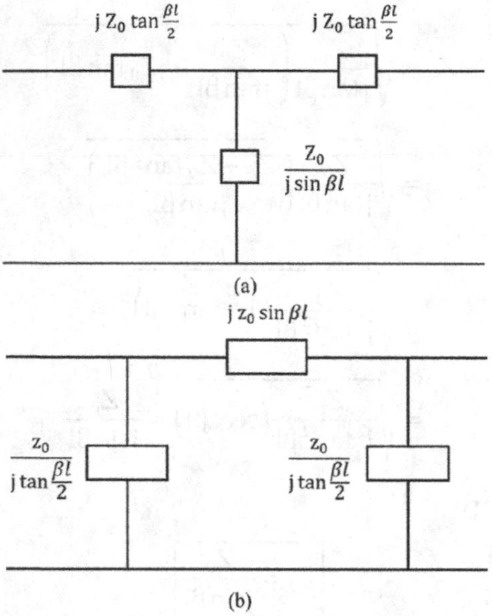

FIGURE 1.3
(a) T-equivalent (b) π-equivalent of a lossless transmission line network.

1.4 T-Network and π-Network

A transmission line of length l and characteristic impedance Z_0 is considered equivalent to the T- or π-network as shown in Figure 1.4. This is obtained by equating open circuit and short circuit impedances of T- and π-networks and also by using equivalent theorem.

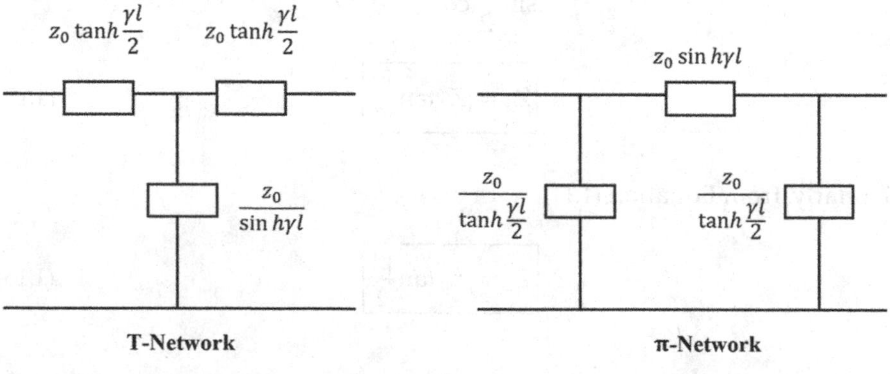

FIGURE 1.4
T- and π-equivalent of general transmission line.

For lossless transmission line, $\alpha = 0$. Therefore, $\gamma = j\beta$. The lossless transmission line network becomes Figure 1.5.

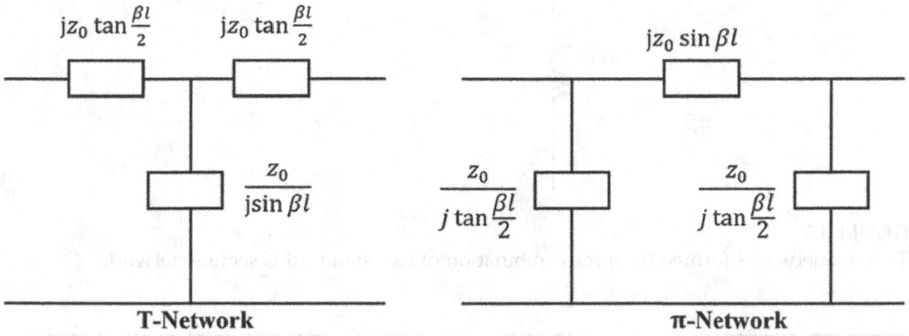

FIGURE 1.5
Lossless transmission line networks.

For all the transmission lines whose length is less than $\lambda_g/4\left(1 < \lambda_g/4\right)$, the lumped element T and π equivalent are given by:

- Low-pass filter (T- and π-type).
- See Figure 1.6.

FIGURE 1.6
Equivalent lumped element circuit of T and π.

1.5 Standard L-Section from Which All Other Network Topologies Are Built

See Figure 1.7.

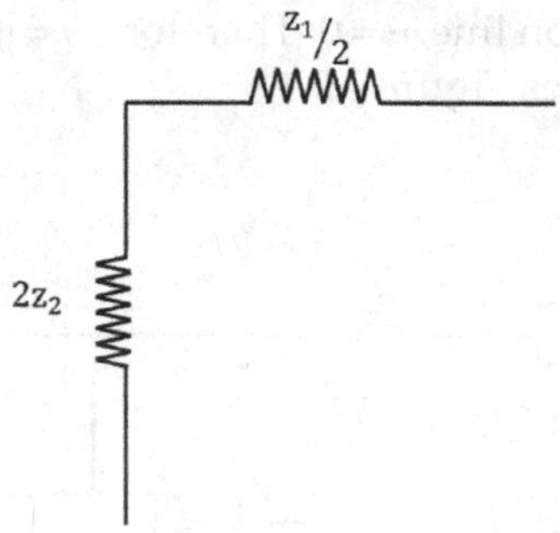

FIGURE 1.7
T- and π-networks formed from the combination of two standard L-section network.

1.6 Standard T-Network and π-Network Formed with Basic L-Section Shown in Figure 1.7

See Figure 1.8.

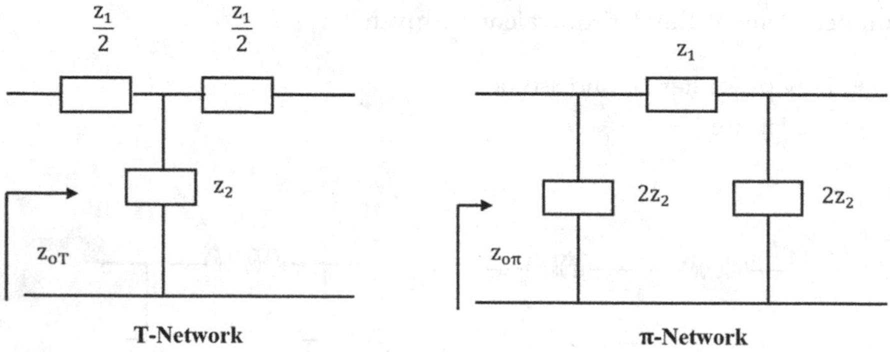

FIGURE 1.8
T- and π-network.

$$\text{For a symmetrical network } Z_i = Z_0 = \sqrt{Z_{oc} Z_{sc}} \tag{1.17}$$

$$\text{For T-network } Z_{oc} = \frac{Z_1}{2} + 2Z_2 \text{ and } Z_{sc} = \frac{Z_1}{2} \tag{1.18}$$

Substituting Equation (1.18) in (1.17)

$$Z_{oT} = \sqrt{\left(\frac{Z_1}{2} + 2Z_2\right)\frac{Z_1}{2}} \tag{1.19}$$

$$\boxed{Z_{oT} = \sqrt{Z_1 Z_2\left(1+\frac{Z_1}{4Z_2}\right)}}. \tag{1.20}$$

Similarly, for π-network

$$Z_{oc} = 2Z_2 \| (Z_1 + Z_2) \text{ and } Z_{sc} = Z_1 \| 2Z_2. \tag{1.21}$$

Substituting Equation (1.21) in (1.17)

$$Z_{o\pi} = \sqrt{Z_1 Z_2 \Big/ \left(1+\frac{Z_1}{4Z_2}\right)}. \tag{1.22}$$

1.7 Relationship between Z_1, Z_2 and Cutoff Frequency (f_c)

From π-network (Figure 1.9)

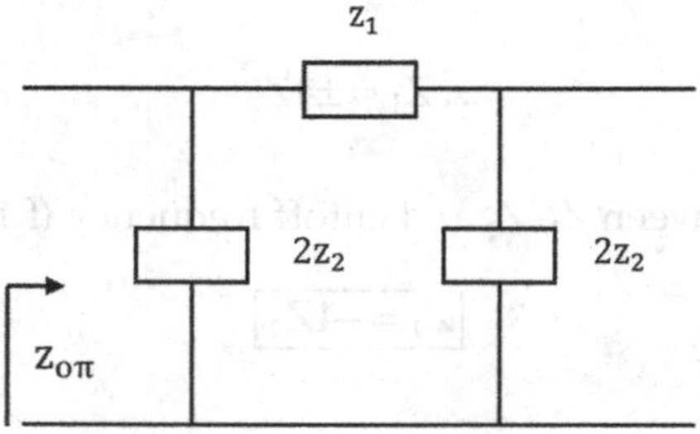

FIGURE 1.9
π-network formed from the combination of two standard L-section network.

Equivalent impendence of this network is given by

$$Z_{o\pi} = \frac{2Z_2}{\frac{Z_1}{2} + 2Z_2} \tag{1.23}$$

Let the cutoff frequency be 3-dB.

$$\frac{4Z_2}{Z_1 + 4Z_2} = \frac{1}{\sqrt{2}}$$

$$\frac{Z_2}{\frac{Z_1}{4} + Z_2} = \frac{1}{\sqrt{2}} \tag{1.24}$$

Tak modulus above equation then we get

$$\left|\frac{Z_2}{\frac{Z_1}{4}+Z_2}\right| = \left|\frac{1}{\sqrt{2}}\right| \Rightarrow \frac{Z_2}{\sqrt{\left(\frac{Z_1}{4}\right)^2 + Z_2{}^2}} = \frac{1}{\sqrt{2}}$$

$$\frac{Z_2{}^2}{\left(\frac{Z_1}{2}\right)^2 + (Z_2)^2} = \frac{1}{2}$$

$$\frac{1}{\left(\frac{1}{4}\right)^4\left(\frac{Z_1}{Z_2}\right)^2 + 1} = \frac{1}{2}$$

$$\left(\frac{1}{4}\right)^2\left(\frac{Z_1}{Z_2}\right)^2 + 1 = 2 \Rightarrow Z_1{}^2 = 16Z_2{}^2$$

$$\therefore Z_1 = \pm 4Z_2$$

The relationship between Z_1, Z_2 and cutoff frequency (f_c) is given by

$$\boxed{Z_1 = -4Z_2} \tag{1.25}$$

If

$Z_1 = L$, $Z_2 = C$, then Equation (1.18) becomes a low-pass filter.
$Z_1 = C$, $Z_2 = L$, then Equation (1.18) becomes a high-pass filter.

1.8 Methods of Realizing L and C

1. Stub method
   Open stub for C
   Short stub for L
2. High-impedance, low-impedance method using Wheeler's curve.

1.8.1 Stub Method

See Figure 1.10.

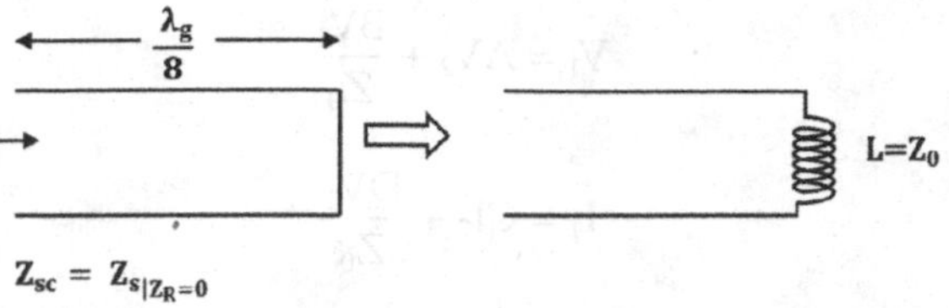

FIGURE 1.10
Short circuited transmission line and its lumped equivalent.

We know that $Z_{sc} = Z_0\left[\dfrac{Z_R\cosh\gamma l + Z_0\sinh\gamma l}{Z_0\cosh\gamma l + Z_R\sinh\gamma l}\right]$

$\therefore\ Z_{sc} = Z_0 \tan h\gamma l$ [$\because$ For Short circuit $Z_R = 0$]

For a lossless transmission line, $\gamma = j\beta$. Hence, the preceding equation becomes

$$Z_{sc} = jZ_0 \cdot \tan\beta l \tag{1.26}$$

See Figure 1.11.

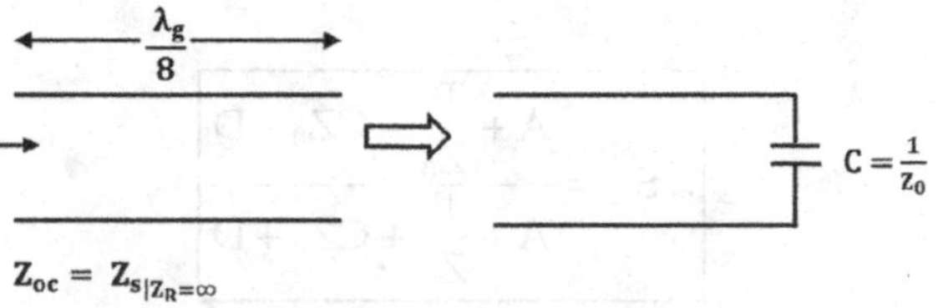

FIGURE 1.11
Open Circuited transmission line and its lumped equivalent.

Substitute $Z_R = \infty$ in Z_S then we get $Z_{oc} = \dfrac{Z_0}{\tanh\gamma l}$ [Open circuit $Z_R = \infty$]

$$\text{For lossless condition}\quad \boxed{Z_{oc} = \frac{Z_0}{j\tan\beta l}} \tag{1.27}$$

1.9 S-Parameters

$$Z_{IN} = \frac{V_1}{I_1} = \frac{AZ_L + B}{CZ_L + D} = \frac{A + \dfrac{B}{Z_0}}{C + \dfrac{D}{Z_0}} \tag{1.28}$$

$$V_1 = AV_2 + \frac{BV_2}{Z_0} \tag{1.29a}$$

$$I_1 = CI_2 + \frac{DV_2}{Z_0} \tag{1.29b}$$

$$P_{IN} = \frac{V_{IN}^-}{V_{IN}^+} = \frac{Z_{IN} - Z_0}{Z_{IN} + Z_0} \tag{1.30}$$

Substitute Equation (1.28) in (1.30) (with $Z_L = Z_0$)

$$P_{IN} = \frac{V_{IN}^-}{V_{IN}^+} = \frac{Z_{IN} - Z_0}{Z_{IN} + Z_0} = \frac{\dfrac{A + \dfrac{B}{Z_0}}{C + \dfrac{D}{Z_0}} - Z_0}{\dfrac{A + \dfrac{B}{Z_0}}{C + \dfrac{D}{Z_0}} + Z_0}$$

$$\boxed{\therefore S_{11} = \frac{A + \dfrac{B}{Z_0} - CZ_0 - D}{A + \dfrac{B}{Z_0} + CZ_0 + D}} \tag{1.31}$$

$$S_{21} = \left.\frac{V_2^-}{V_1^-}\right|_{V_2^+ = 0} = \frac{V_2}{V_1}(1 + S_{11})$$

From Equation (1.29a), $\dfrac{V_2}{V_1} = \dfrac{1}{A + \dfrac{B}{Z_0}}$

$$S_{21} = \frac{1}{A + \dfrac{B}{Z_0}}\left(1 + \frac{A + \dfrac{B}{Z_0} - CZ_0 - D}{A + \dfrac{B}{Z_0} + CZ_0 + D}\right)$$

$$= \frac{1}{A + \dfrac{B}{Z_0}}\left(\frac{A + \dfrac{B}{Z_0} + CZ_0 + D + A + \dfrac{B}{Z_0} - CZ_0 - D}{A + \dfrac{B}{Z_0} + CZ_0 + D}\right) = \frac{2}{A + \dfrac{B}{Z_0} + CZ_0 + D}$$

$$\boxed{\therefore S_{21} = \frac{2}{A + \dfrac{B}{Z_0} + CZ_0 + D}} \tag{1.32}$$

$$Z_{OUT} = \frac{V_2}{I_2} = \frac{DZ_0 + B}{CZ_0 + A} = \frac{D + \dfrac{B}{Z_0}}{C + \dfrac{A}{Z_0}}$$

$$\therefore S_{22} = \frac{Z_{OUT} - Z_0}{Z_{OUT} + Z_0} = \frac{\dfrac{D + \frac{B}{Z_0}}{C + \frac{A}{Z_0}} - Z_0}{\dfrac{D + \frac{B}{Z_0}}{C + \frac{A}{Z_0}} + Z_0} = \frac{D + \dfrac{B}{Z_0} - CZ_0 - A}{D + \dfrac{B}{Z_0} + CZ_0 + A} = \frac{-A + \dfrac{B}{Z_0} - CZ_0 + D}{A + \dfrac{B}{Z_0} + CZ_0 + D}$$

$$\boxed{\therefore S_{22} = \frac{-A + \dfrac{B}{Z_0} - CZ_0 + D}{A + \dfrac{B}{Z_0} + CZ_0 + D}} \tag{1.33}$$

$$S_{12} = \left.\frac{V_1^-}{V_1^+}\right|_{V_1^+=0} = \frac{V_1}{V_2}(1 + S_{22}) \tag{1.34}$$

Therefore, Equation (1.29a) becomes

$$V_1 = AV_2 - \frac{BV_2}{Z_0} \Rightarrow \frac{V_1}{V_2} = A - \frac{B}{Z_0}$$

$$\therefore S_{12} = A - \frac{B}{Z_0}\left(1 + \frac{-A + \dfrac{B}{Z_0} - CZ_0 - D}{A + \dfrac{B}{Z_0} + CZ_0 + D}\right)$$

Substituting Z_0 value in this equation (Figure 1.12),

$$S_{12} = A - \frac{B\left(C+\frac{A}{Z_0}\right)}{D+\frac{B}{Z_0}}\left(\frac{D+\frac{B}{Z_0}}{A+\frac{B}{Z_0}+CZ_0+D}\right)\times 2[\because Z_0 = Z_{OUT}]$$

$$= \frac{AD+\frac{AB}{Z_0}-BC-\frac{AB}{Z_0}}{D+\frac{B}{Z_0}}\left(\frac{D+\frac{B}{Z_0}}{A+\frac{B}{Z_0}+CZ_0+D}\right)\times 2$$

$$= 2\left[\frac{AD-BC}{A+\frac{B}{Z_0}+CZ_0+D}\right]$$

$$\therefore S_{12} = 2\left[\frac{AD-BC}{A+\frac{B}{Z_0}+CZ_0+D}\right] \tag{1.35}$$

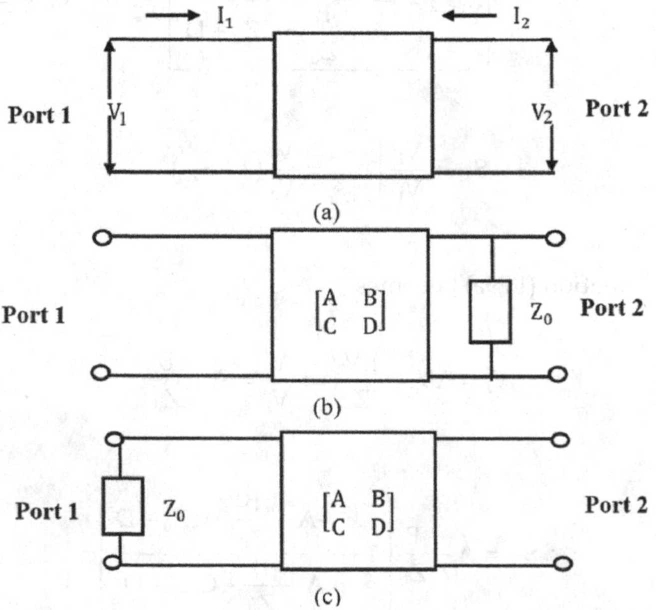

FIGURE 1.12
ABCD parameters.

Therefore, the S-parameters are given by

$$\begin{bmatrix} S_{11} & S_{12} \\ S_{21} & S_{22} \end{bmatrix} = \frac{1}{\Delta}\begin{bmatrix} A+\frac{B}{Z_0}-CZ_0-D & 2(AD-BC) \\ 2 & -A+\frac{B}{Z_0}-CZ_0+D \end{bmatrix},$$

where $\Delta = A+\frac{B}{Z_0}+CZ_0+D$

$$[Y]=\frac{1}{Z_0}\begin{bmatrix} \frac{D}{B} & \frac{BC-AD}{B} \\ -\frac{1}{B} & \frac{A}{B} \end{bmatrix}$$

$$\begin{bmatrix} A & B \\ C & D \end{bmatrix} = \frac{1}{Y_{21}}\begin{bmatrix} -Y_{22} & -1 \\ Y_{12}Y_{21}-Y_{11}Y_{22} & -Y_{11} \end{bmatrix}$$

Transmission matrix

$$\begin{bmatrix} T_{11} & T_{12} \\ T_{21} & T_{22} \end{bmatrix} = \begin{bmatrix} \frac{1}{S_{21}} & -\frac{S_{22}}{S_{21}} \\ \frac{S_{11}}{S_{21}} & \frac{S_{12}S_{21}-S_{11}S_{22}}{S_{21}} \end{bmatrix}$$

$$\begin{bmatrix} S_{11} & S_{12} \\ S_{21} & S_{22} \end{bmatrix} = \begin{bmatrix} \frac{T_{21}}{T_{11}} & -\frac{T_{11}T_{22}-T_{12}T_{21}}{T_{11}} \\ \frac{1}{T_{11}} & -\frac{T_{12}}{T_{11}} \end{bmatrix}$$

$$Z_{IN} = \frac{1+\Gamma_{in}}{1-\Gamma_{in}}.$$

From S_{11}, S_{12}, S_{21}, S_{22},

$$\left.\begin{aligned} A &= \frac{(1+S_{11})(1-S_{22})+S_{12}S_{21}}{2S_{21}} \\ B &= \frac{Z_0(1+S_{11})(1-S_{22})-S_{12}S_{21}}{2S_{21}} \\ C &= \frac{1}{Z_0}\frac{(1-S_{11})(1-S_{22})-S_{12}S_{21}}{2S_{21}} \\ D &= \frac{(1-S_{11})(1+S_{22})+S_{12}S_{21}}{2S_{21}} \end{aligned}\right\} \tag{1.36}$$

1.10 ABCD Parameters

A set of parameters useful for transmission line networks are the ABCD parameters or the chain parameters. The ABCD parameter is defined in terms of the voltages and currents as shown in Figure 1.13.

1.10.1 ABCD Parameters General Form

$$V_1 = AV_2 - BI_2$$

$$I_1 = AV_2 - DI_2$$

$$\begin{bmatrix} V_1 \\ I_1 \end{bmatrix} = \begin{bmatrix} A & B \\ C & D \end{bmatrix} \begin{bmatrix} V_2 \\ -I_2 \end{bmatrix} \quad (1.37)$$

1.10.2 Series Impedance

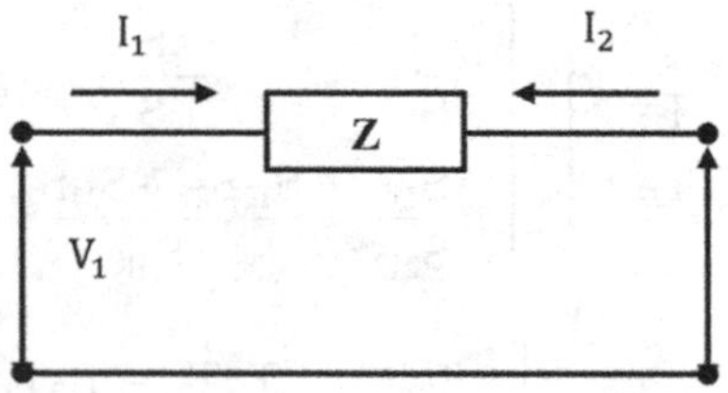

FIGURE 1.13
Series impedance Z.

$$V_1 = V_2 - I_2 Z$$

$$I_1 = -I_2$$

$$\begin{bmatrix} V_1 \\ I_1 \end{bmatrix} = \begin{bmatrix} 1 & Z \\ 0 & 1 \end{bmatrix} \begin{bmatrix} V_2 \\ -I_2 \end{bmatrix}$$

$$\begin{bmatrix} A & B \\ C & D \end{bmatrix}_Z = \begin{bmatrix} 1 & Z \\ 0 & 1 \end{bmatrix} \quad (1.38)$$

where

A - reverse voltage gain

B - reverse transfer impedance

C - reverse transfer admittance

D - reverse current gain

1.10.3 Conversion of ABCD Parameters of Series Impedance to S-Parameters

From Equations (1.31)–(1.33) and (1.35)

$$\left.\begin{aligned} S_{11} &= \frac{A + \frac{B}{Z_0} - CZ_0 - D}{\Delta} & S_{12} &= \frac{2\,(AD - BC)}{\Delta} \\ S_{21} &= \frac{2}{\Delta} & S_{22} &= \frac{-A + \frac{B}{Z_0} - CZ_0 + D}{\Delta} \end{aligned}\right\} \quad (1.39)$$

$$\text{where} \quad \Delta = A + \frac{B}{Z_0} + CZ_0 + D$$

$$S_{11} = \frac{A + \frac{B}{Z_0} - CZ_0 - D}{\Delta} = \frac{Z}{Z + 2Z_0} [\because \text{ substitute ABCD values from 1.38}] \quad (1.40)$$

$$S_{12} = \frac{2(AD - BC)}{\Delta} = \frac{2Z_0}{Z + 2Z_0} \quad (1.41)$$

$$S_{21} = \frac{2}{\Delta} = \frac{2Z_0}{Z + 2Z_0} \quad (1.42)$$

$$S_{22} = \frac{-A + \frac{B}{Z_0} - CZ_0 + D}{\Delta} = \frac{Z}{Z + 2Z_0} \quad (1.43)$$

$$\therefore [S]_Z = \begin{bmatrix} \frac{Z}{Z + 2Z_0} & \frac{2Z_0}{Z + 2Z_0} \\ \frac{2Z_0}{Z + 2Z_0} & \frac{Z}{Z + 2Z_0} \end{bmatrix} \quad (1.44)$$

1.10.4 Shunt Admittance

See Figure 1.14.

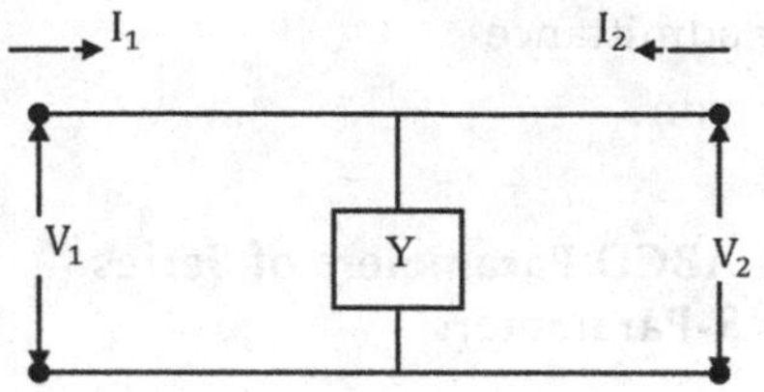

FIGURE 1.14
Shunt admittance Y.

$$V_1 = V_2$$

$$I_1 = V_2Y - I_2$$

$$\begin{bmatrix} V_1 \\ I_1 \end{bmatrix} = \begin{bmatrix} 1 & 0 \\ Y & 1 \end{bmatrix} \begin{bmatrix} V_2 \\ -I_2 \end{bmatrix}$$

$$\boxed{\begin{bmatrix} A & B \\ C & D \end{bmatrix}_Y = \begin{bmatrix} 1 & 0 \\ Y & 1 \end{bmatrix}} \tag{1.45}$$

1.10.5 Conversion of ABCD Parameters of Shunt Admittance to S-Parameters

$$\left.\begin{aligned} S_{11} &= \frac{A + BY_0 - c/Y_0 - D}{\Delta} & S_{12} &= \frac{2(AD - BC)}{\Delta} \\ S_{21} &= \frac{2}{\Delta} & S_{22} &= \frac{-A + BY_0 - c/Y_0 + D}{\Delta} \end{aligned}\right\} \tag{1.46}$$

where $\Delta = A + BY_0 + c/Y_0 + D$

$$S_{11} = \frac{A + BY_0 - c/Y_0 - D}{\Delta} = \frac{-Y}{Y + 2Y_0} \quad [\because \text{substitute ABCD values from 1.45}] \tag{1.47}$$

$$S_{12} = \frac{2(AD - BC)}{\Delta} = \frac{2Y_0}{Y + 2Y_0} \tag{1.48}$$

$$S_{21} = \frac{2}{\Delta} = \frac{2Y_0}{Y + 2Y_0} \tag{1.49}$$

$$S_{22} = \frac{-A + BY_0 - c/Y_0 + D}{\Delta} = \frac{-Y}{Y + 2Y_0} \tag{1.50}$$

$$\therefore [S]_Y = \begin{bmatrix} \frac{-Y}{Y+2Y_0} & \frac{2Y_0}{Y+2Y_0} \\ \frac{2Y_0}{Y+2Y_0} & \frac{-Y}{Y+2Y_0} \end{bmatrix} \text{(or)} \begin{bmatrix} \frac{-YZ_0}{2+YZ_0} & \frac{2}{2+YZ_0} \\ \frac{2}{2+YZ_0} & \frac{-YZ_0}{2+YZ_0} \end{bmatrix} \tag{1.51}$$

where Z_0 = characteristic impedance.

1.10.6 Series Impedance Cascade with a Shunt Admittance

See Figure 1.15.

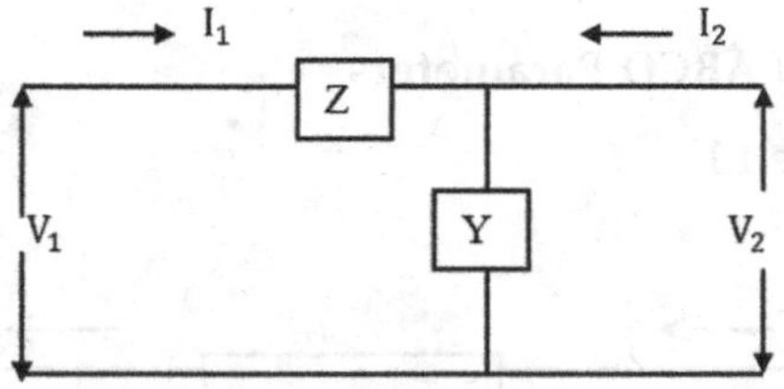

FIGURE 1.15
Series impedance Z cascade with a shunt admittance Y.

$$\begin{bmatrix} A & B \\ C & D \end{bmatrix} = \begin{bmatrix} A_1 & B_1 \\ C_1 & D_1 \end{bmatrix} \begin{bmatrix} A_2 & B_2 \\ C_2 & D_2 \end{bmatrix}$$

$$= \begin{bmatrix} 1 & Z \\ 0 & 1 \end{bmatrix} \begin{bmatrix} 1 & 0 \\ Y & 1 \end{bmatrix} = \begin{bmatrix} 1+ZY & Z \\ Y & 1 \end{bmatrix}$$

$$\begin{bmatrix} A & B \\ C & D \end{bmatrix}_{ZY} = \begin{bmatrix} 1+ZY & Z \\ Y & 1 \end{bmatrix} \tag{1.52}$$

1.10.7 Transformer

See Figure 1.16.

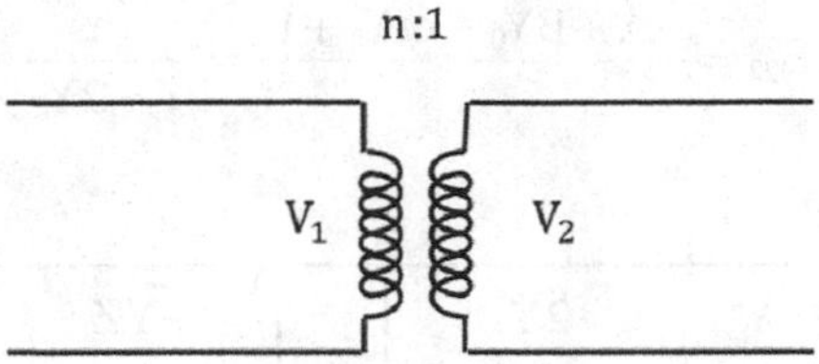

FIGURE 1.16
An ideal transformer of turns ratio n:1.

$$V_1 = V_2 n$$

$$I_1 = -I_2$$

$$\begin{bmatrix} V_1 \\ I_1 \end{bmatrix} = \begin{bmatrix} n & 0 \\ 0 & 1/n \end{bmatrix} \begin{bmatrix} V_2 \\ -I_2 \end{bmatrix} \tag{1.53}$$

1.10.8 Properties of ABCD Parameters

See Figure 1.17; Table 1.1.

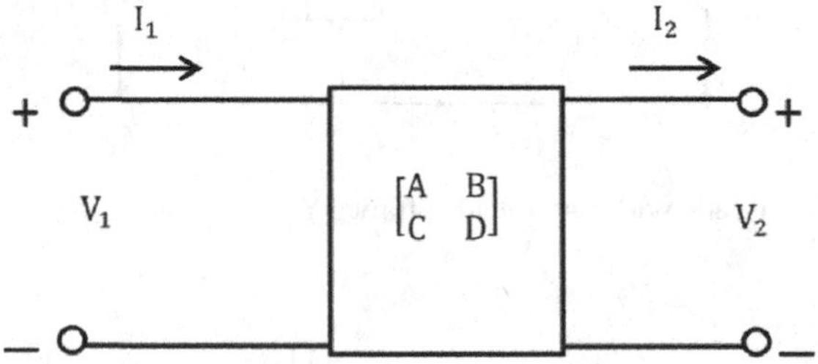

FIGURE 1.17
ABCD network.

$$\left.\begin{array}{c} \begin{bmatrix} V_1 \\ I_1 \end{bmatrix} = \begin{bmatrix} A & B \\ C & D \end{bmatrix} \begin{bmatrix} V_2 \\ I_2 \end{bmatrix} \\ AD - BC = 1 \text{ (for a Reciprocal Network)} \\ A = D \text{ (for a Symmetrical Network)} \\ A \text{ and } D \text{ are real (for a Lossless Network)} \\ B \text{ and } C \text{ are Imaginary} \\ Z_{I1} = \sqrt{\dfrac{AB}{CD}} \quad Z_{I2} = \sqrt{\dfrac{BD}{AC}} \quad Z_{input} = \dfrac{AZ_L + B}{CZ_L + D} \end{array}\right\} \tag{1.54}$$

TABLE 1.1
Summary – Properties of [Z], [S], and [ABCD]

Property	[Z]	[S]	[ABCD]
Reciprocal	$Z_{12} = Z_{21}$	$S_{12} = S_{21}$	AD – BC=1
Symmetrical	$Z_{11} = Z_{22}$	$S_{11} = S_{22}$	A = D
Lossless	[Z] Imag	$[S]^t[S]^* = [U]$	A, D Real B, C Imaginary

1.11 Two-Port Networks Matched on Image and Iteration Basics

See Figures 1.18–1.20.

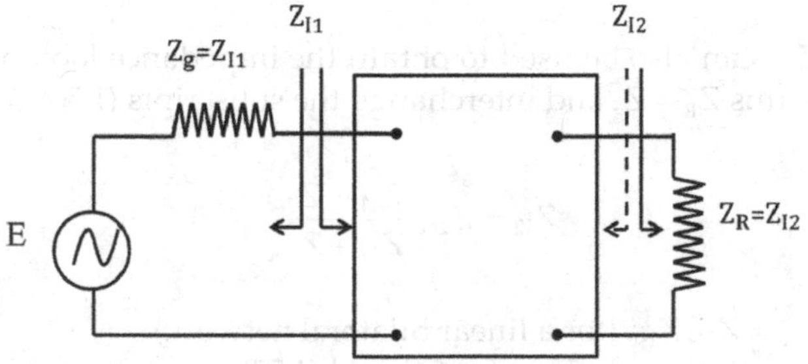

FIGURE 1.18
Four-terminal network terminated on an image basis.

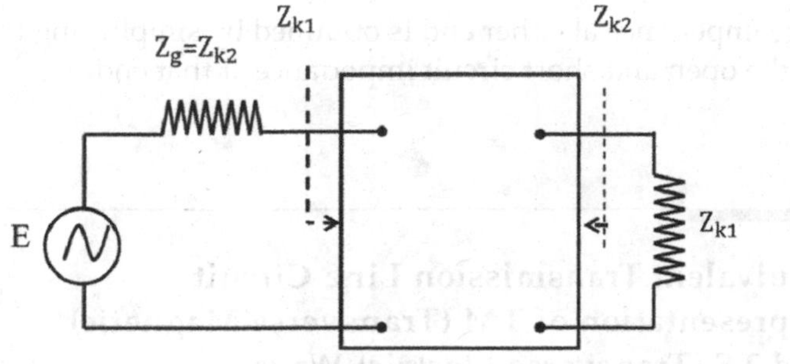

FIGURE 1.19
Four-terminal network terminated in its iterative impedances.

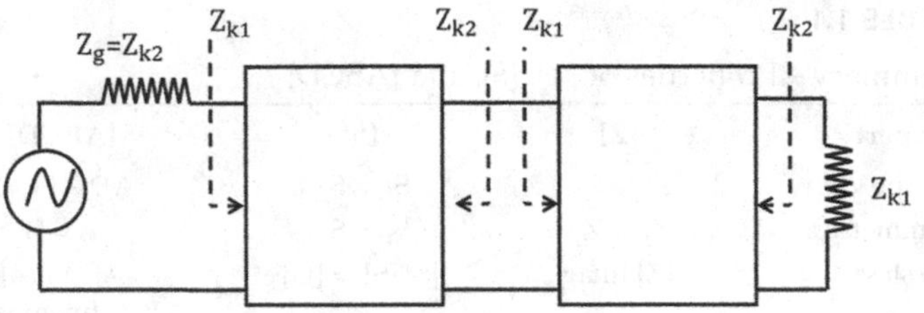

FIGURE 1.20
Changing line networks terminated in its iterative impedances at both ends.

$$\boxed{Z_S = Z_{1oc}\frac{Z_R + Z_{2sc}}{Z_R + Z_{2oc}}} \tag{1.55}$$

Substitute $Z_R = Z_{I2}$ and $Z_S = Z_{I1}$ in (1.55)

$$\boxed{\therefore Z_{I1} = Z_{1oc}\frac{Z_{I2} + Z_{2sc}}{Z_{I2} + Z_{2oc}}} \tag{1.56}$$

Equation (1.56) can also be used to obtain the impedance looking to the left at port 2. For this $Z_R = Z_g$ and interchange the subscripts (1.56)

$$Z_{I2} = Z_{2oc}\frac{Z_{I1} + Z_{2sc}}{Z_{I1} + Z_{2oc}} \tag{1.57}$$

Using $Z_{1oc}Z_{2sc} = Z_{2oc}Z_{1sc}$ (for a linear bilateral network)

Simultaneously, the solution of (1.56) and (1.57) is as follows:

$$\left.\begin{aligned} Z_{I1} &= \sqrt{Z_{1oc}Z_{1sc}} \\ Z_{I2} &= \sqrt{Z_{2oc}Z_{2sc}} \end{aligned}\right\} \tag{1.58}$$

The image impedance at either end is obtained by simplifying the geometry between the open and short circuit impedance at that end.

1.12 Equivalent Transmission Line Circuit Representation of TM (Transverse Magnetic) and TE (Transverse Electric) Waves

See Figure 1.21.

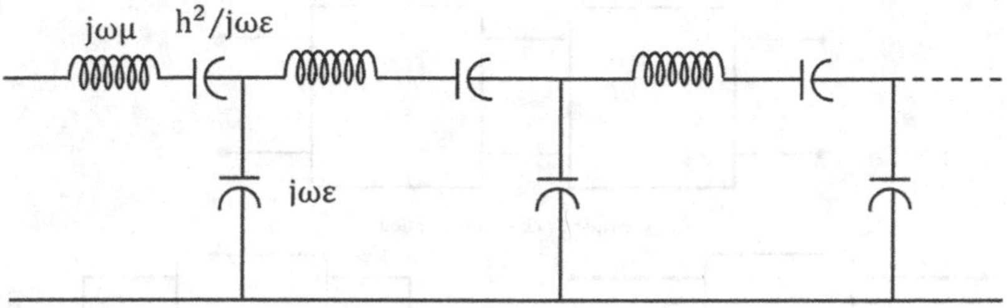

FIGURE 1.21
Equivalent transmission line circuit representation of TM waves.

The characteristic impedance of TM (Figure 1.22)

$$Z_0(TM) = \sqrt{\frac{Z}{Y}} = \sqrt{\frac{j\omega\mu + h^2/j\omega\varepsilon}{j\omega\varepsilon}} = \sqrt{\frac{\mu}{\varepsilon}}\sqrt{1-\frac{\omega_c^{\ 2}}{\omega^2}} \tag{1.59}$$

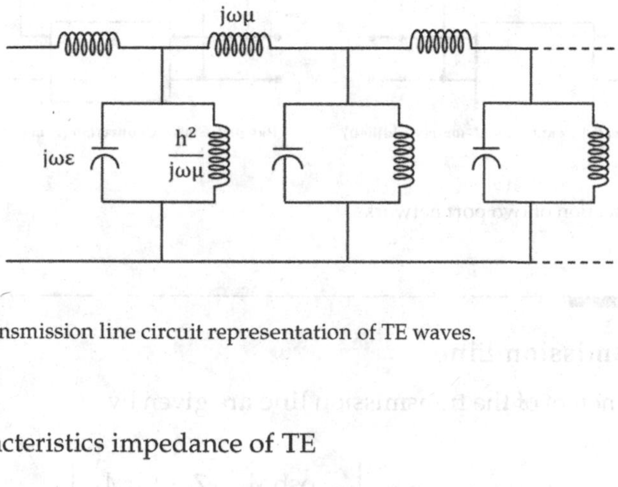

FIGURE 1.22
Equivalent transmission line circuit representation of TE waves.

The characteristics impedance of TE

$$Z_0(TM) = \sqrt{\frac{j\omega\mu}{j\omega\varepsilon + h^2/j\omega\mu}} = \sqrt{\frac{\mu}{\varepsilon}}\sqrt{\frac{1}{1-\left(\frac{\omega_c^{\ 2}}{\omega^2}\right)}} \tag{1.60}$$

1.13 Basic Interconnection of the Two-Port Network

See Figure 1.23.

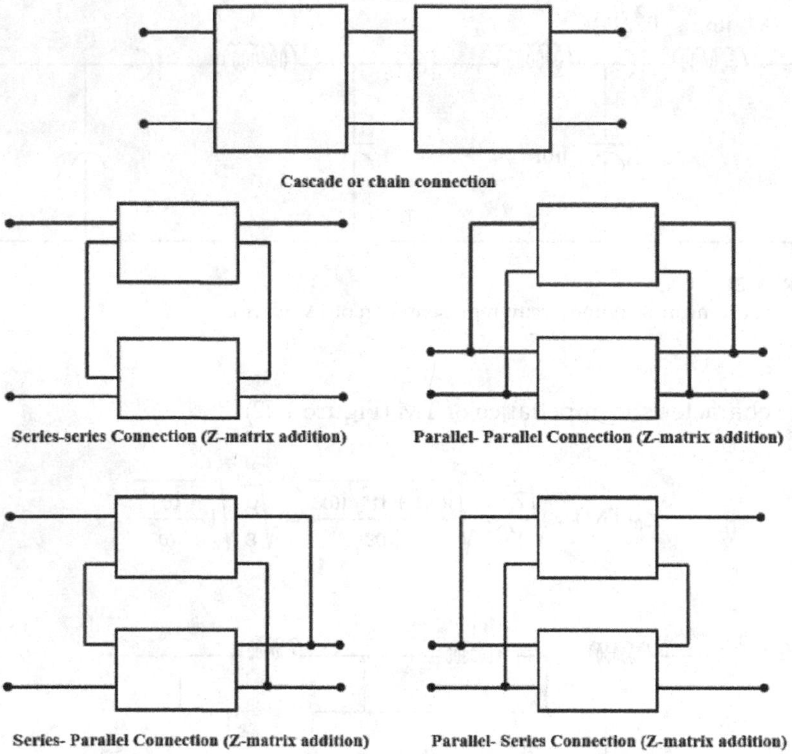

FIGURE 1.23
Basic interconnection of two port networks.

1.14 Transmission Line

ABCD parameters of the transmission line are given by

$$\begin{bmatrix} A & B \\ C & D \end{bmatrix}_{TL} = \begin{bmatrix} \cosh\gamma l & Z_0 \sinh\gamma l \\ \dfrac{\sinh\gamma l}{Z_0} & \cosh\gamma l \end{bmatrix} \tag{1.61}$$

For lossless condition, Equation (1.61) becomes

$$\boxed{\begin{bmatrix} A & B \\ C & D \end{bmatrix}_{TL} = \begin{bmatrix} \cos\beta l & jZ_0 \sin\beta l \\ \dfrac{j\sin\beta l}{Z_0} & \cos\beta l \end{bmatrix}} \tag{1.62}$$

For an ideal transmission line, the S-parameter matrix is given as

$$[S]_{Matrix} = \begin{bmatrix} 0 & e^{-j\beta l} \\ e^{-j\beta l} & 0 \end{bmatrix} \tag{1.63}$$

1.15 Effective ABCD Parameters

See Figure 1.24; Table 1.2.

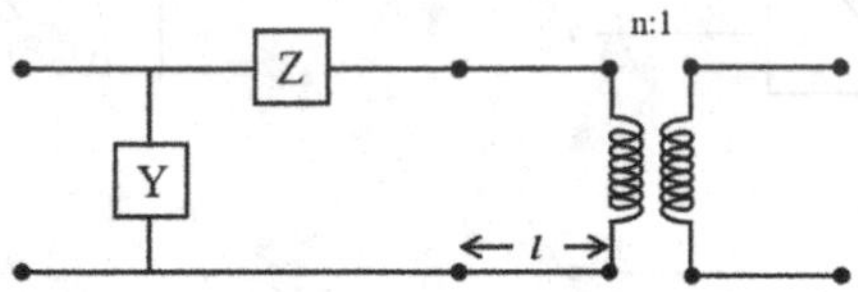

FIGURE 1.24
Effective ABCD parameters.

$$\begin{bmatrix} A & B \\ C & D \end{bmatrix}_{effective}$$

$$= \begin{bmatrix} 1 & 0 \\ Y & 1 \end{bmatrix} \begin{bmatrix} 1 & Z \\ 0 & 1 \end{bmatrix} \begin{bmatrix} \cos\beta l & jZ_0 \sin\beta l \\ \dfrac{j\sin\beta l}{Z_0} & \cos\beta l \end{bmatrix} \begin{bmatrix} n & 0 \\ 0 & 1/n \end{bmatrix} \tag{1.64}$$

1.16 Conversion of ABCD Parameters of the Transformer into S-Parameters

$$\left.\begin{aligned} S_{11} &= \frac{A + \dfrac{B}{Z_0} - CZ_0 - D}{\Delta} & S_{12} &= \frac{2(AD - BC)}{\Delta} \\ S_{21} &= \frac{2}{\Delta} & S_{22} &= \frac{-A + \dfrac{B}{Z_0} - CZ_0 + D}{\Delta} \end{aligned}\right\} \tag{1.65}$$

where $\Delta = A + \dfrac{B}{Z_0} + CZ_0 + D$

TABLE 1.2
Transmission Matrix Parameters of Some Basic Circuit Elements

Network	$\begin{bmatrix} A & B \\ C & D \end{bmatrix}$
Transmission Line Section	
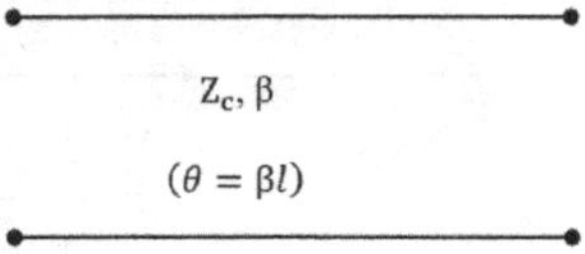	$\begin{bmatrix} \cos\theta & jZ_c \sin\theta \\ \dfrac{j\sin\theta}{Z_c} & \cos\theta \end{bmatrix}$
Series Impedance	
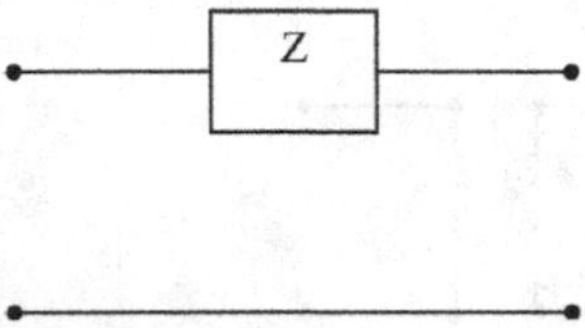	$\begin{bmatrix} 1 & Z \\ 0 & 1 \end{bmatrix}$
Shunt Admittance	
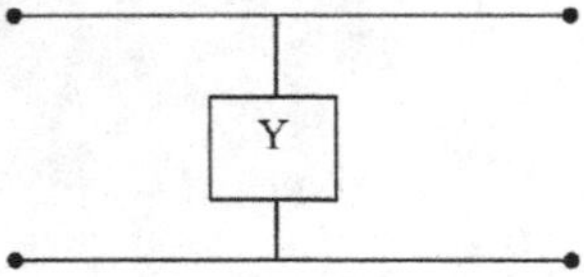	$\begin{bmatrix} 1 & 0 \\ Y & 1 \end{bmatrix}$
Transformer	
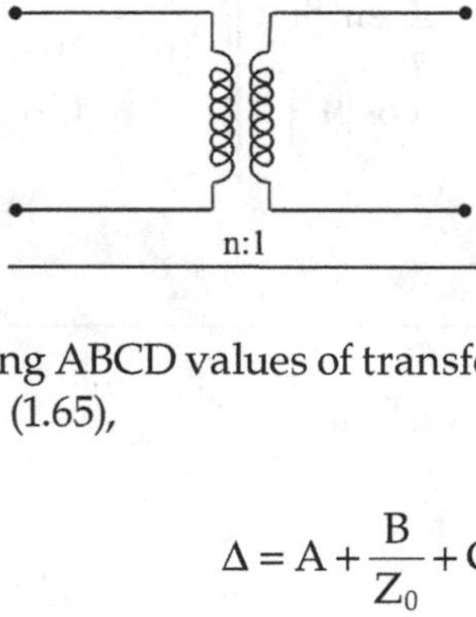	$\begin{bmatrix} n & 0 \\ 0 & 1/n \end{bmatrix}$

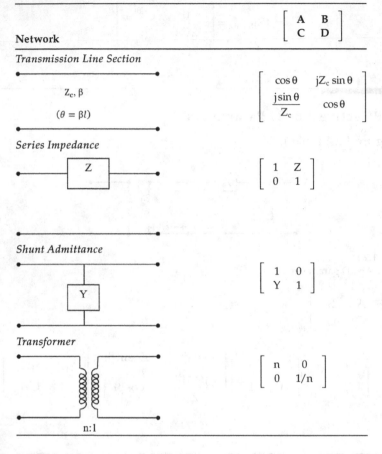

Obtaining ABCD values of transformer from Equation (1.53) and substituting them in (1.65),

$$\Delta = A + \frac{B}{Z_0} + CZ_0 + D = n + \frac{1}{n} = \frac{n^2+1}{n} \tag{1.66}$$

$$S_{11} = \frac{n - 1/n}{n^2 + 1/n} = \frac{n^2-1}{n} \times \frac{n}{n^2+1} = \frac{n^2-1}{n^2+1}$$

$$S_{12} = \frac{2\,(n \times 1/n)}{n^2 + 1/n} = \frac{2n}{n^2+1}$$

$$S_{21} = \frac{2}{\Delta} = \frac{2}{n^2 + 1/n} = \frac{2n}{n^2+1}$$

$$S_{22} = \frac{-n+1/n}{n^2+1/n} = \frac{-n^2+1}{n} \times \frac{n}{n^2+1} = \frac{-n^2+1}{n^2+1}$$

$$\therefore [S]_{Transformer} = \begin{bmatrix} \frac{n^2-1}{n^2+1} & \frac{2n}{n^2+1} \\ \frac{2n}{n^2+1} & \frac{-n^2+1}{n^2+1} \end{bmatrix} \quad (1.67)$$

1.17 Unit Element (UE)

See Figure 1.25.

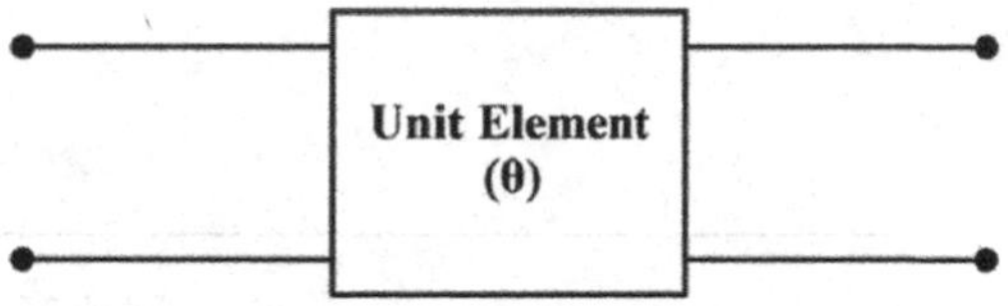

FIGURE 1.25
Unit element.

$$\begin{bmatrix} A & B \\ C & D \end{bmatrix}_{TL} = \begin{bmatrix} \cos\theta & jZ_0 \sin\theta \\ \frac{j\sin\theta}{Z_0} & \cos\theta \end{bmatrix}$$

where $\theta = \beta l$

$$\begin{bmatrix} A & B \\ C & D \end{bmatrix}_{UE} = \cos\theta \begin{bmatrix} 1 & jZ_0 \tan\theta \\ \frac{j\tan\theta}{Z_0} & 1 \end{bmatrix}$$

Substituting $S = j\tan\theta$ in the preceding equation we get

$$\begin{bmatrix} A & B \\ C & D \end{bmatrix}_{UE} = \cos\theta \begin{bmatrix} 1 & SZ_0 \\ \frac{S}{Z_0} & 1 \end{bmatrix}$$

$$= \frac{1}{\sec\theta} \begin{bmatrix} 1 & SZ_0 \\ \frac{S}{Z_0} & 1 \end{bmatrix}$$

$$= \frac{1}{\sqrt{1+\tan^2\theta}} \begin{bmatrix} 1 & SZ_0 \\ \dfrac{S}{Z_0} & 1 \end{bmatrix}$$

$$= \frac{1}{\sqrt{1-S^2}} \begin{bmatrix} 1 & SZ_0 \\ \dfrac{S}{Z_0} & 1 \end{bmatrix} \quad \left[\because \begin{matrix} S^2 = j^2 \tan^2\theta \\ S^2 = -\tan^2\theta \end{matrix} \right]$$

$$\boxed{\begin{bmatrix} A & B \\ C & D \end{bmatrix}_{UE} = \frac{1}{\sqrt{1-S^2}} \begin{bmatrix} 1 & SZ_0 \\ \dfrac{S}{Z_0} & 1 \end{bmatrix}} \tag{1.68}$$

1.18 K-Inverter (Impedance Inverter)

See Figure 1.26.

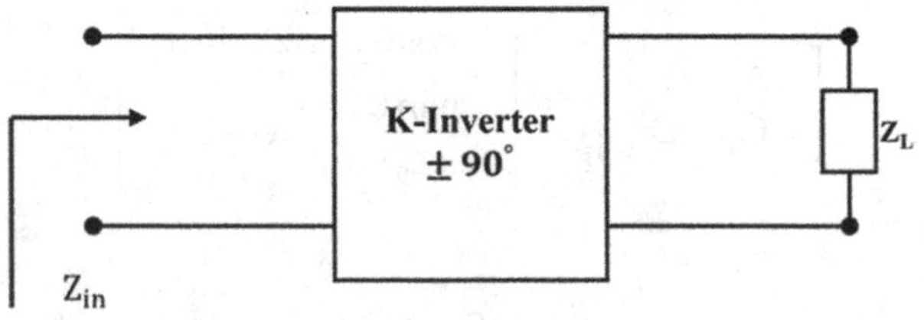

FIGURE 1.26
K-inverter.

The ABCD parameters for transmission line are given as

$$\begin{bmatrix} A & B \\ C & D \end{bmatrix}_{TL} = \begin{bmatrix} \cos\beta l & jZ_0 \sin\beta l \\ \dfrac{j\sin\beta l}{Z_0} & \cos\beta l \end{bmatrix} \tag{1.69}$$

where $\beta = \dfrac{2\Pi}{\lambda_g}$ and $l = \dfrac{\lambda_g}{4}$ [Length of the inverter]

$$\therefore \beta l = \frac{\pi}{2} \tag{1.70}$$

Substituting Equation (1.70) in (1.69)

$$\begin{bmatrix} A & B \\ C & D \end{bmatrix}_K = \begin{bmatrix} 0 & jZ_0 \\ \dfrac{j}{Z_0} & 0 \end{bmatrix}$$

For K-inverter, $Z_0 = K$. Therefore, the abovementioned equation becomes

$$\boxed{\begin{bmatrix} A & B \\ C & D \end{bmatrix}_K = \begin{bmatrix} 0 & jK \\ \dfrac{j}{K} & 0 \end{bmatrix}} \tag{1.71}$$

1.19 J-Inverter (Admittance Inverter)

See Figure 1.27.

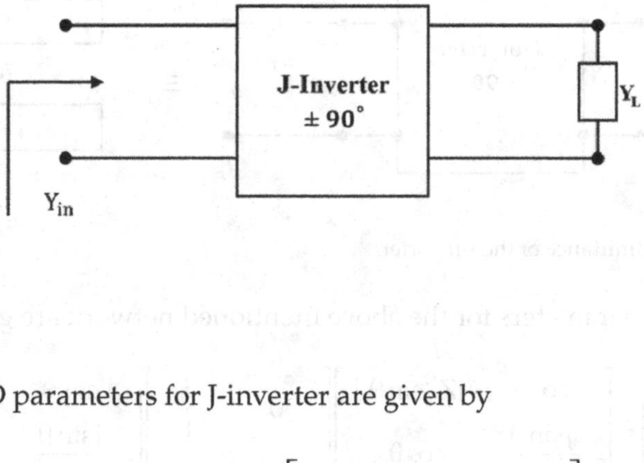

FIGURE 1.27
J-inverter.

The ABCD parameters for J-inverter are given by

$$\begin{bmatrix} A & B \\ C & D \end{bmatrix}_{TL} = \begin{bmatrix} \cos\beta l & \dfrac{j\sin\beta l}{Y_0} \\ jY_0 \sin\beta l & \cos\beta l \end{bmatrix} \tag{1.72}$$

where $\beta = \dfrac{2\Pi}{\lambda_g}$ and $l = \dfrac{\lambda_g}{4}$ [Length of the inverter]

$$\therefore \beta l = \frac{\pi}{2} \tag{1.73}$$

Substituting Equation (1.73) in (1.72)

$$\begin{bmatrix} A & B \\ C & D \end{bmatrix}_J = \begin{bmatrix} 0 & \frac{j}{Y_0} \\ jY_0 & 0 \end{bmatrix}$$

For J-inverter, $Y_0 = J$. Therefore, the given equation becomes

$$\boxed{\begin{bmatrix} A & B \\ C & D \end{bmatrix}_J = \begin{bmatrix} 0 & \frac{j}{J} \\ jJ & 0 \end{bmatrix}} \tag{1.74}$$

1.20 Analysis of Odd Mode and Even Mode

See Figure 1.28.

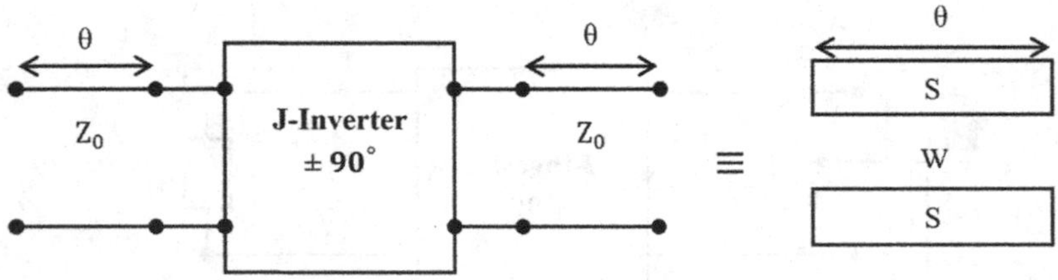

FIGURE 1.28
Characteristic admittance of the J-inverter.

The ABCD parameters for the above mentioned network are given by

$$\begin{bmatrix} A & B \\ C & D \end{bmatrix} = \begin{bmatrix} \cos\theta & jZ_0\sin\theta \\ \frac{j\sin\theta}{Z_0} & \cos\theta \end{bmatrix} \begin{bmatrix} 0 & -\frac{j}{J} \\ -jJ & 0 \end{bmatrix} \begin{bmatrix} \cos\theta & jZ_0\sin\theta \\ \frac{j\sin\theta}{Z_0} & \cos\theta \end{bmatrix} \tag{1.75}$$

$$\begin{bmatrix} A & B \\ C & D \end{bmatrix} = \begin{bmatrix} \left(JZ_0 + \frac{1}{JZ_0}\right)\sin\theta\cos\theta & j\left(JZ_0{}^2\sin^2\theta - \frac{\cos^2\theta}{J}\right) \\ j\left(\frac{1}{JZ_0{}^2}\sin^2\theta - J\cos^2\theta\right) & \left(JZ_0 + \frac{1}{JZ_0}\right)\sin\theta\cos\theta \end{bmatrix} \tag{1.76}$$

The ABCD parameters of the admittance inverter were obtained by considering quarter wavelength of transmission line.

From coupled mode theory

$$Z_i = \frac{1}{2}[Z_{oe} - Z_{oo}] \tag{1.77}$$

$$\cos\beta = A = \left(\frac{Z_{0e} + Z_{oo}}{Z_{oe} - Z_{oo}}\right)\cos\theta \tag{1.78}$$

From ABCD parameter $Z_i = \sqrt{\frac{AB}{CD}} = \sqrt{\frac{B}{C}}$ [If network is symmetrical A = D]

(1.79)

Getting B and C values from Equation (1.76) and substituting them in Equation (1.79)

$$Z_i = \sqrt{\frac{B}{C}} = \sqrt{\frac{j\left(JZ_0{}^2\sin\theta - \frac{\cos^2\theta}{J}\right)}{j\left(\frac{1}{JZ_0{}^2\sin^2\theta} - J\cos^2\theta\right)}} \tag{1.80}$$

where $\theta = \beta l$, $\beta = \frac{2\pi}{\lambda_g}$, and $l = \frac{\lambda_g}{4}$ [Length of the inverter]

$$\boxed{\therefore \theta = \frac{\pi}{2}} \tag{1.81}$$

Substituting Equation (1.81) in (1.80)

$$Z_i = \sqrt{\frac{B}{C}} = \sqrt{\frac{j\left(JZ_0{}^2\right)}{j\left(\frac{1}{JZ_0{}^2}\right)}} = \sqrt{J^2 Z_0{}^4}$$

$$Z_i = JZ_0{}^2 \tag{1.82}$$

Substituting Equation (1.82) in (1.77)

$$JZ_0{}^2 = \frac{1}{2}[Z_{oe} - Z_{oo}]$$

$$Z_{oe} - Z_{oo} = 2JZ_0^{\,2} \tag{1.83}$$

Obtaining "A" value from Equation (1.76) and substituting them in Equation (1.78) we get

$$\left(JZ_0 + \frac{1}{JZ_0}\right)\sin\theta\cos\theta = \left(\frac{Z_{0e} + Z_{oo}}{Z_{oe} - Z_{oo}}\right)\cos\theta$$

$$\left(JZ_0 + \frac{1}{JZ_0}\right) = \left(\frac{Z_{0e} + Z_{oo}}{Z_{oe} - Z_{oo}}\right)\left[\because \begin{matrix} \theta = \beta l = \dfrac{\pi}{2} \\ \sin\dfrac{\pi}{2} = 1 \end{matrix}\right]$$

$$\left(\frac{J^2 Z_0^{\,2} + 1}{JZ_0}\right) = \left(\frac{Z_{0e} + Z_{oo}}{2JZ_0^{\,2}}\right)\text{[by 1.83]}$$

$$Z_{0e} + Z_{0o} = 2J^2 Z_0^{\,3} + 2Z_0 \tag{1.84}$$

Adding Equations (1.83) and (1.84) we get

$$2Z_{0e} = 2JZ_0^{\,2} + 2J^2 Z_0^{\,3} + 2Z_0$$

$$2Z_{0e} = 2\left(Z_0 + J + Z_0^{\,2} + J^2 Z_0^{\,3}\right)$$

$$\boxed{Z_{0e} = Z_0\left[1 + JZ_0 + (JZ_0)^2\right]} \tag{1.85}$$

Subtracting Equation (1.83) from (1.84)

$$\boxed{Z_{0o} = Z_0\left[1 - JZ_0 + (JZ_0)^2\right]} \tag{1.86}$$

Similarly, for K-inverter

$$\boxed{\begin{aligned} Z_{0e} &= Z_0\left[1 + \frac{K}{Z_0} + \left(\frac{k}{Z_0}\right)^2\right] \\ Z_{0e} &= Z_0\left[1 - \frac{K}{Z_0} + \left(\frac{k}{Z_0}\right)^2\right] \end{aligned}} \tag{1.87}$$

1.21 Kuroda's Identities

1.21.1 Introduction

Kuroda's Identities provide equivalent two-port networks. Both the equivalent networks have the same transmission, admittance/impedance, and scattering matrices. Thus, one- and two-port networks can be replaced with their equivalent without changing the characteristics of the network.

- The four Kuroda's Identities are redundant transmission line sections to achieve a more practical microwave filter implementation by performing any of the following operations:
  i. Physically separate transmission line stubs
  ii. Transform series stubs into stubs or vice versa
  iii. Change impractical characteristic impedance into more realizable ones.

- The additional transmission lines are called unit elements or UEs and are $\frac{\lambda}{8}$ long at cutoff frequency (ω_c).
- The UEs are commensurate with the stubs obtained by Richard's transformation from the prototype design.
- The inductors and capacitors represented by the Kuroda's Identities are short circuit and open circuit stubs.

1.21.2 First Kuroda Identity

See Figure 1.29.

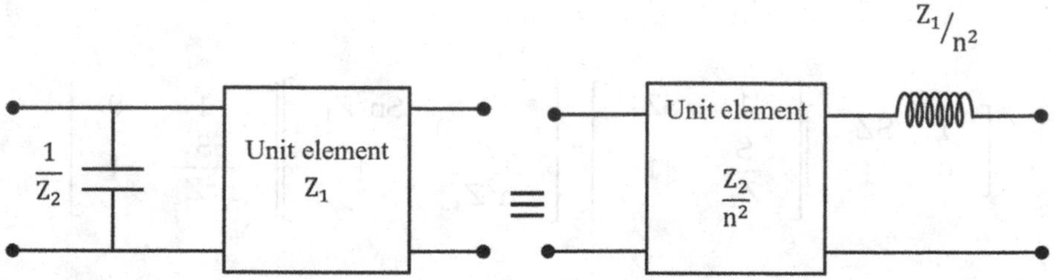

FIGURE 1.29
First Kuroda identity.

Matrix for LHS (Left Hand Side) circuit

$$\begin{bmatrix} 1 & 0 \\ \dfrac{S}{Z_2} & 1 \end{bmatrix} \frac{1}{\sqrt{1-S^2}} \begin{bmatrix} 1 & SZ_1 \\ \dfrac{S}{Z_1} & 1 \end{bmatrix} = \frac{1}{\sqrt{1-S^2}} \begin{bmatrix} 1 & SZ_1 \\ S\left(\dfrac{1}{Z_1}+\dfrac{1}{Z_2}\right) & S^2\dfrac{Z_1}{Z_2}+1 \end{bmatrix} \tag{1.88}$$

Matrix for RHS (Right Hand Side) circuit

$$\frac{1}{\sqrt{1-S^2}} \begin{bmatrix} 1 & S\dfrac{Z_2}{n^2} \\ \dfrac{Sn^2}{Z_2} & 1 \end{bmatrix} \begin{bmatrix} 1 & S\dfrac{Z_1}{n^2} \\ 0 & 1 \end{bmatrix} = \frac{1}{\sqrt{1-S^2}} \begin{bmatrix} 1 & S\left(\dfrac{Z_1}{n^2}+\dfrac{Z_2}{n^2}\right) \\ \dfrac{Sn^2}{Z_2} & S^2\dfrac{Z_1}{Z_2}+1 \end{bmatrix} \tag{1.89}$$

Equations (1.88) and (1.89) are equal if $n^2 = 1 + \dfrac{Z_2}{Z_1}$.

1.21.3 Second Kuroda Identity

See Figure 1.30.

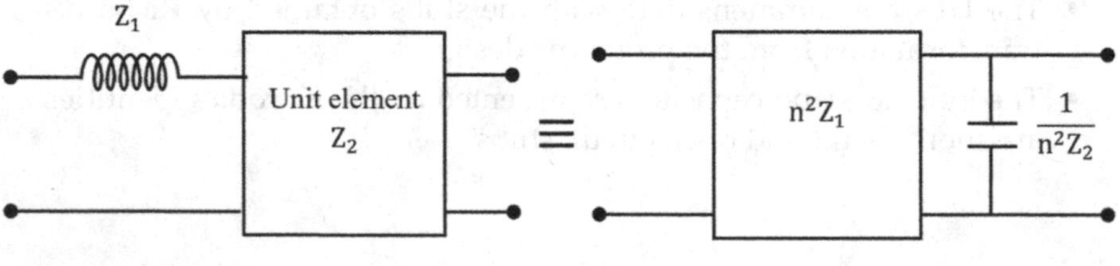

FIGURE 1.30
Second Kuroda identity.

$$\begin{bmatrix} 1 & SZ_1 \\ 0 & 1 \end{bmatrix} \begin{bmatrix} 1 & SZ_2 \\ \dfrac{S}{Z_2} & 1 \end{bmatrix} = \begin{bmatrix} 1 & Sn^2Z_1 \\ \dfrac{S}{n^2Z_1} & 1 \end{bmatrix} \begin{bmatrix} 1 & 0 \\ \dfrac{S}{n^2Z_2} & 1 \end{bmatrix}$$

$$\begin{bmatrix} 1+S^2\dfrac{Z_1}{Z_2} & S(Z_2+Z_1) \\ \dfrac{S}{Z_2} & 1 \end{bmatrix} = \begin{bmatrix} 1+\dfrac{S^2n^2Z_1}{n^2Z_2} & Sn^2Z_1 \\ \dfrac{S}{n^2Z_1}+\dfrac{S}{n^2Z_2} & 1 \end{bmatrix} \tag{1.90}$$

$$\begin{bmatrix} 1+S^2\dfrac{Z_1}{Z_2} & S(Z_2+Z_1) \\ \dfrac{S}{Z_2} & 1 \end{bmatrix} = \begin{bmatrix} 1+\dfrac{S^2Z_1}{Z_2} & Sn^2Z_1 \\ \dfrac{S}{n^2}\left(\dfrac{1}{Z_1}+\dfrac{1}{Z_2}\right) & 1 \end{bmatrix} \tag{1.91}$$

From (1.90) and (1.91)

$$S(Z_2+Z_1) = Sn^2Z_1 \Rightarrow \boxed{n^2 = 1+\frac{Z_2}{Z_1}} \tag{1.92}$$

1.21.4 Fourth Kuroda Identity

See Figure 1.31.

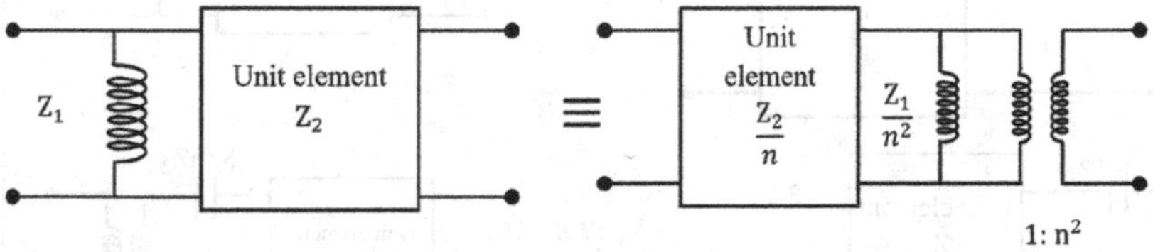

FIGURE 1.31
Fourth Kuroda identity.

$$\text{LHS}\begin{bmatrix} 1 & 0 \\ \dfrac{1}{SZ_1} & 1 \end{bmatrix}\frac{1}{\sqrt{1-S^2}}\begin{bmatrix} 1 & SZ_2 \\ \dfrac{S}{Z_2} & 1 \end{bmatrix}$$

$$= \frac{1}{\sqrt{1-S^2}}\begin{bmatrix} 1 & SZ_2 \\ \dfrac{1}{SZ_1}+\dfrac{S}{Z_2} & \dfrac{Z_2}{Z_1}+1 \end{bmatrix} \tag{1.93}$$

$$\text{RHS}\ \frac{1}{\sqrt{1-S^2}}\begin{bmatrix} 1 & S\dfrac{Z_2}{n^2} \\ \dfrac{Sn^2}{Z_2} & 1 \end{bmatrix}\begin{bmatrix} 1 & 0 \\ \dfrac{n^2}{SZ_2} & 1 \end{bmatrix}\begin{bmatrix} 1/n^2 & 0 \\ 0 & n^2 \end{bmatrix}$$

$$= \frac{1}{\sqrt{1-S^2}}\begin{bmatrix} 1/n^2+\dfrac{Z_2}{Z_1n^2} & SZ_1 \\ \dfrac{1}{SZ_1}+\dfrac{S}{Z_2} & n^2 \end{bmatrix} \tag{1.94}$$

Equations (1.93) and (1.94) are identical if $n^2 = 1 + \frac{Z_2}{Z_1}$ (Tables 1.3 and 1.4).

TABLE 1.3
The Four Kuroda's Identities

Initial Circuit | **Kuroda's Identity**

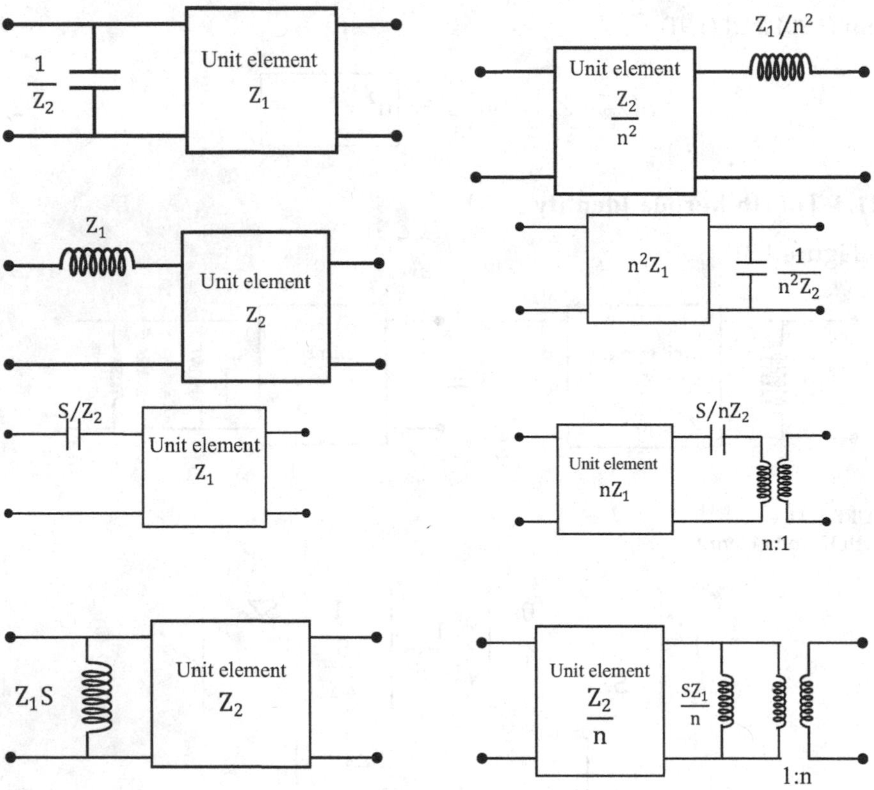

Where $n^2 = 1 + \frac{Z_2}{Z_1}$

TABLE 1.4

Equivalent Circuit for Impedance and Admittance Inverter

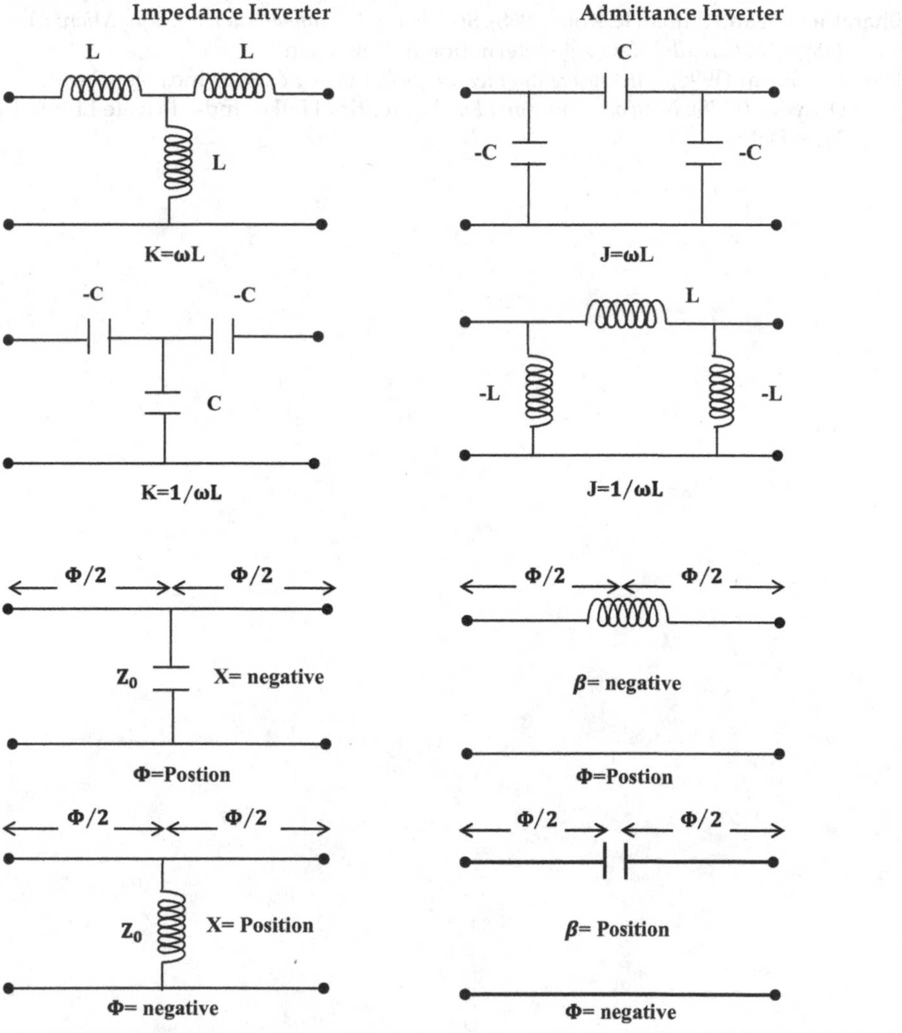

1.21.5 Conclusion

The sending impedance equation, ABCD parameters, K-J Inverters, Kurodas identities. LC realization by 'stepped impedance method' and 'stub method', coupled mode analysis, state of the art of the Smith chart, and the various lengths of guided wavelengths (lambdas) are important in the MIC components design

Bibliography

Bharathi Bhat and Shiban K. Koul (1968). *Stripline-Like Transmission Lines for Microwave Integrated Circuits*, New Age International, New Delhi.

David M. Pozar (1990). *Microwave Engineering*, John Wiley & Sons Inc., New York.

John D. Ryder (1978). *Networks Lines and Fields*, Prentice Hall of India Private Limited, New Delhi.

2

Planar Transmission Lines

The electromagnetic spectrum. See Figure 2.1.

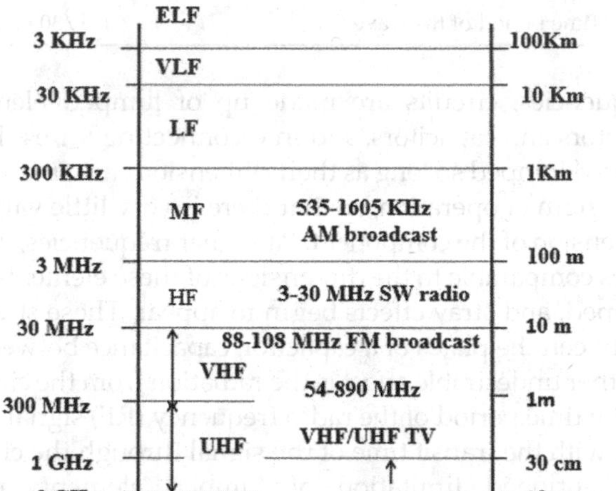

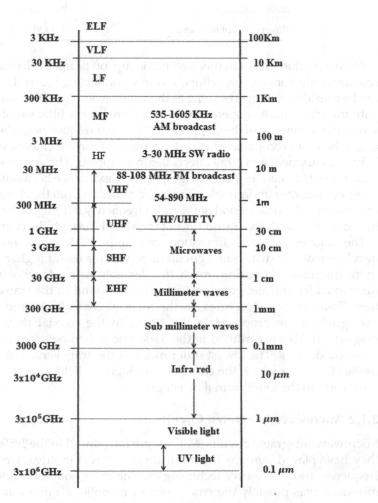

FIGURE 2.1
The electromagnetic spectrum.

2.1 Microwave Theory and Circuits

2.1.1 Introduction to Microwaves

"Microwaves" represent electromagnetic waves in the frequency range from 1 to 30 GHz.

Frequency (f)	1–30 GHz
Wavelength $\lambda_0 = \frac{v_0}{f}$	30–1 cm
Time period of the wave (T)	1–1/30 ns

At low frequencies, circuits are made up of lumped elements, namely, resistors, inductors and capacitors, and interconnecting wires. These elements can be considered lumped so long as their dimensions are small in comparison with the wavelength of operation, so that there is very little variation in phase across the dimension of the component. At higher frequencies, when the wavelength becomes comparable to the dimensions of these elements, they no more behave as lumped, and stray effects begin to appear. These strays include the inductance between the plates of a capacitor, capacitance between the turns of a coil, etc. Another undesirable effect is the radiation from the circuit. Radiation occurs when the time period of the radio frequency (RF) signal becomes small in comparison with the transit time of the signal through the circuit.

The abovementioned limitation of lumped elements prompted the development of a distributed circuit approach for use at higher frequencies. In the distributed circuit approach, the elements are made up of transmission lines in which the electromagnetic fields are bound in the transverse direction. Transmission structures, in the form of coaxial line and rectangular waveguide, were proposed in the 1890s. In the coaxial line, the electromagnetic fields are confined to the dielectric region between the inner and outer conductors. The lowest order mode is the transverse electromagnetic mode (TEM), for which the electric and magnetic field lines lie in the plane transverse to the longitudinal direction.

2.1.2 Microwave Integrated Circuits

Microwave integrated circuits (MICs) were introduced in the 1950s. Since then they have played perhaps the most important role in advancing the radius frequency and microwave technologies. The most noticeable and important milestone was possibly the emergence of monolithic microwave integrated circuits (MMICs). The progress of MICs would not have been possible without the advances of solid-state devices and planar transmission lines.

Planar transmission lines refer to transmission lines that consist of conducting strips printed on the surface of the transmission line substrates. These structures are the backbone of MICs and represent an important and interesting research topic for many microwave engineers. Along with the advances in MICs and planar transmission lines, numerous analysis methods for microwave and millimeter-wave passive structures in general, and planar transmission lines in particular, have been developed in response to the need for accurate analysis and design of MICs. These analysis methods have in turn helped further investigation and development of new planar transmission lines.

Useful and commonly used techniques for analyzing microwave planar transmission lines in particular and passive structures in general:

1. Green's function
2. Conformal mapping
3. Variational
4. Special domain
5. Mode matching (Figure 2.2)

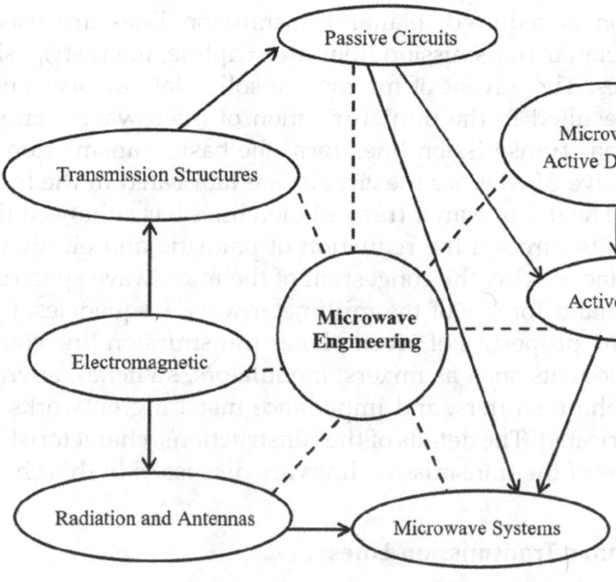

FIGURE 2.2
Typical contents of the microwave engineering program.

2.1.3 Introduction to Planar Transmission Lines

The transmission of a signal or energy from a source to some distant point requires the use of a transmission medium, generally called a transmission line. Transmission lines are either lumped lines or distributed lines. The circuit at low frequencies is fabricated by using lumped components such as resistors, inductors, capacitors, and transistors, or vacuum tubes. At audio frequencies, these lumped elements are small in comparison with the wavelength of operation. However, at higher frequencies, where the wavelengths are comparable to the dimensions of these components, stray elements begin to appear. Typical strays include the inductance between the plates of a capacitor, capacitance between the turns of an inductor coil, the inductance of the interconnecting conductive wires or strips, and the capacitance between them. Another undesirable effect at radio frequencies is that the circuit radiates energy and is easily electrically disturbed by any nearby objects. Hence, these effects become significant at microwave frequencies.

To circumvent these limitations of the lumped circuits, the distributed circuit approach that is, transmission lines, namely, coaxial line, and rectangular waveguide were developed for use at microwave frequencies. The emphasis was shifted from the currents flowing inside the conductors to the electromagnetic waves propagating in the dielectric medium inside these transmission lines. The available media was compatible with these devices. The answer was realized in the form of planar transmission lines.

At frequencies ranging from 1 to 300 GHz and where large circuit bandwidth or miniaturization is required, planar transmission lines are used. The commonly used planar transmission lines are stripline, microstrip, slotline, and coplanar lines. The advent of microwave solid-state devices and MIC technology has resulted in the miniaturization of microwave components and systems. Planar transmission lines form the basic transmission media for MICs irrespective of whether the circuits are fabricated in the hybrid or monolithic form. The use of planar transmission lines has enhanced the performance of circuits through the reduction of parasitic and elimination of superfluous interfaces. Also, the congestion of the microwave spectrum has increased the demand for use of the millimeter-wave frequencies. By carefully exploiting the properties of given planar transmission line configuration, useful components such as mixers, modulators, switches, attenuators, filters, couplers, phase shifters, and impedance matching networks can be designed and fabricated. The details of the constructions, characteristics, and important features of the transmission lines are discussed in this chapter.

2.1.4 Various Planar Transmission Lines

The parallel plate line with infinite plates and propagating a TEM wave is a form of transmission line, which has no losses. This form of transmission line is only theoretical and not practical due to its infinite size and the difficulties of supporting the plates. Recently, a line using finite plates and on an intervening medium to support the plates has found growing importance. Thus, two possible configurations exist and are called *stripline* and *microstrip*.

For a frequency above the microwave range, conventional planar transmission line structures suffer from fabrication problems, due to which some modifications of the basic stripline and microstrip configurations are essential. Some of the modified versions are suspended stripline, suspended microstrip, inverted microstrip, coplanar stripline (CPS), coplanar waveguide (CPW), and its variants.

The fundamental mode of propagation in conventional transmission lines, such as the two-wire line and the coaxial line, and the balanced stripline is essentially a TEM. The planar transmission lines such as microstrip, slotline, and CPW (Coplanar Waveguide) are quasi-TEM lines.

At frequencies above the microwave range, it is necessary to use thinner substrates in microstrips to have propagation in the dominant mode (quasi-TEM mode) at the exclusion of higher order modes. The width of the strip conductor becomes progressively narrower, demanding increasingly stringent dimensional tolerances during the fabrication, and in addition, the line losses tend to be excessive. These problems are circumvented to a large extent in the suspended substrate lines by incorporating an air gap between the substrate and the ground plane.

The analysis of the planar transmission line is required to determine its characteristic impedance (Z_0) and effective dielectric constant (ε_{eff}). The analysis includes dynamic analysis and static analysis. Dynamic analysis permits one to calculate the hybrid fundamental mode (quasi-TEM) and often the hybrid higher order modes by taking into account all field components. This yields frequency-dependent values for phase and characteristic impedance, which uses the solutions of the Helmholtz wave equations.

In static analysis, the solution of Laplace's equation subject to the boundary conditions leads to the static electric potential values in the structure. Then, the capacitances in the structure with and without the substrates are defined. From the characteristic impedance, effective dielectric constant and the phase velocity are obtained in the quasi-static limit.

In the inhomogeneous medium, calculations are made numerically by using numerical methods such as finite difference method (FDM), finite element method (FEM), spectral domain method, conformal mapping method, variational method, transverse transmission line (TTL) method, boundary element method, integral equation method, and mode matching method. The error of static approximation as compared with dynamic approximation increases as frequency increases.

2.1.5 Parameters for Selection of Transmission Structure

Among the available various types of guides for applications in millimeter-wave circuits, the selection of a particular transmission structure for a given frequency range depends upon the following considerations:

i. Low loss or high unloaded Q
ii. Low radiation loss
iii. Low dispersion
iv. Maximum achievable bandwidth
v. Ease of bonding solid-state devices in the case of active components

vi. Ease of integration
vii. Adequate power handling
viii. Ease of fabrication and low cost of manufacture.

2.1.6 Calculation of Z_0, v_{ph}, and ε_{eff}

If a quasi-TEM wave is assumed to propagate along a transmission line, the wave impedance and phase velocity may be determined from the two capacitance values C and C_a, with and without dielectric filling, respectively.

For a lossless transmission line, R = 0 and G = 0, that is, in the quasi-static limit, and hence, the characteristic impedance is

$$Z_0 = \sqrt{\frac{R + j\omega L}{G + j\omega C}} = \sqrt{\frac{L}{C}} \tag{2.1}$$

The propagation constant is $\gamma = \sqrt{(R + j\omega L)(G + j\omega C)} = j\omega\sqrt{LC}$. Hence, the phase constant is $\beta = \omega\sqrt{LC}$ and its phase velocity is $v_{ph} = \frac{\omega}{\beta} = \sqrt{\frac{1}{LC}}$, where R, L, C, and G represent series resistance, series inductance, shunt capacitance, and shunt conductance, respectively.

When the dielectric material is air throughout, the phase velocity must be equal to the velocity of light (v_0). If "C_a" is the capacitance in that case, then

$$v_0 = \frac{1}{\sqrt{LC_a}} \tag{2.2}$$

Hence, the line inductance, $L = \frac{1}{v_0^2 C_a}$.

Since both dielectrics have the permeability of free space, the inductance value is unaffected by the presence of dielectric. Hence, the characteristic impedance becomes

$$Z_0 = \sqrt{\frac{L}{C}} = \frac{1}{v_0\sqrt{CC_a}} \tag{2.3}$$

and phase velocity,

$$v_{ph} = \sqrt{\frac{1}{LC}} = v_0\sqrt{\frac{C_a}{C}} = \frac{v_0}{\sqrt{C/C_a}} = \frac{v_0}{\sqrt{\varepsilon_{eff}}} \tag{2.4}$$

Here, "C" is the capacitance per unit length of the structure, "C_a" is the capacitance per unit length of the same structure with all dielectrics replaced by air, and ε_{eff} is the effective permittivity.

2.2 Planar Transmission Lines and Microwave Integrated Circuits

1. Planar transmission lines are essential components of MICs.
2. It has been used to realize many circuit functions such as BALUNS, FIBERS, HYBRIDS, and COUPLERS, as well as simply to carry signals.

Commonly used planar transmission lines:

1. Microstriplines
2. Striplines
3. Suspended striplines
4. Finlines
5. Slot lines
6. Inverted microstripline
7. Coplanar waveguides
8. Coplanar strips.

Beyond certain frequencies, conducting wires are of no use due to skin depth. Waveguides (circular, ridged, and rectangular) come into use. As the frequency increases beyond GHz, MICs play a dominant role. The backbone of MICs is the planar transmission line. Planar transmission lines include stripline, microstrip, slot line, coplanar waveguide, coplanar strips, finline, substrate integrated waveguide, and their variants. Suspended microstrip, inverted microstrip, trapped microstrip, image guides, dielectric integrated guides, and suspended stripline are few of the variants of planar transmission lines. The cross sections, field lines, and the modes of each transmission line need to be understood to make the design procedure easier. Stripline supports the pure TEM mode, microstrip supports the quasi-TEM mode, slot line supports the non-TEM mode, and finline supports the hybrid mode. The design of MIC components is in terms of guide wavelength (from "zero to one" wavelength, there are infinite wavelengths, and few of them have become standard wavelengths), and hence the design formulae remain the same; what changes is the effective dielectric constant. For a homogenous transmission line like stripline, the relative dielectric constant is to be used, and for all other transmission lines, effective dielectric constant needs to be taken into consideration. This is due to "dispersion." If it is mm-wave application, CPW family and finlines are the ideal choices. In RF microelectromechanical systems (MEMS). BioMEMS also CPW and Microstrip play dominant roles. MICROWAVE INTEGRATED CIRCUIT Design if and only if the simple BASICS of planar transmission lines are understood.

2.3 Stripline

The first planar transmission line, called the "strip transmission line" was proposed by Barrett and Barnes. The cross-sectional view of the *stripline* is shown in Figure 2.3a. It consists of two parallel "ground planes" spaced at distance "h" apart and with central strip conductor of width "'w" and thickness "t." It is centrally positioned within the dielectric material with a relative permittivity of about 2 or 3. This appears as a flattened coaxial transmission line. Propagation in stripline is somewhat similar to that in coaxial cable wherein the fields are confined between the outer and inner conductors. Thus, if the distance between the plates is small compared to a wavelength, losses are low due to the absence of higher order modes, and propagation is essentially in the TEM mode with the electric and magnetic field configurations shown in Figure 2.3b.

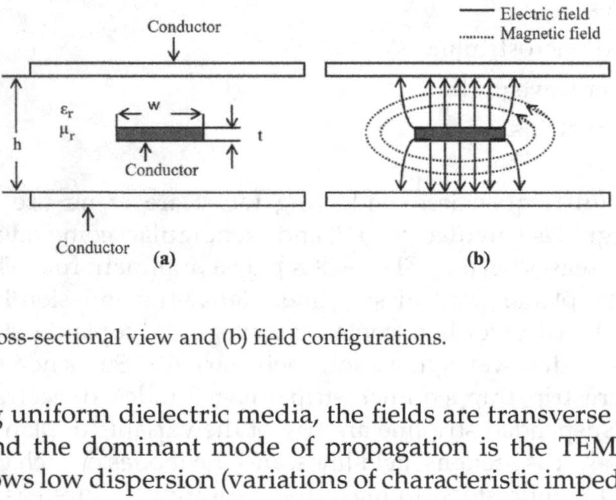

FIGURE 2.3
Stripline: (a) cross-sectional view and (b) field configurations.

Assuming uniform dielectric media, the fields are transverse electric and magnetic and the dominant mode of propagation is the TEM mode. The stripline shows low dispersion (variations of characteristic impedance Z_0 and relative permittivity ε_r with frequency). The homogeneous stripline is ideal for passive integrated circuits such as filters, couplers, and other hybrids.

For a lossless stripline, Z_0 may be written as

$$Z_0 = \frac{1}{C v_{ph}} \tag{2.5}$$

where C is the value of capacitance per unit length and ν_{ph} is the phase velocity.

The phase velocity, characteristic impedance, and wavelength of propagation λ_g of the wave along the stripline are given by

$$v_{ph} = \frac{v_0}{\sqrt{\varepsilon_r}} \tag{2.6}$$

$$Z_0 = \frac{Z_a}{\sqrt{\varepsilon_r}} \tag{2.7}$$

$$\lambda_g = \frac{\lambda_0}{\sqrt{\varepsilon_r}} \tag{2.8}$$

where ν_0 is the velocity of electromagnetic waves in free space and ε_r is the relative dielectric constant of the dielectric medium.

The characteristic impedance of Z_0 may be written as (accuracy within 1%)

$$\left.\begin{aligned} Z_0 &= \frac{30}{\sqrt{\varepsilon_r}} \ln\left\{2\left(\frac{1+\sqrt{k}}{1-\sqrt{k}}\right)\right\}\Omega, \quad \text{for } w/h \le 0.5 \\ Z_0 &= \frac{30\pi^2}{\sqrt{\varepsilon_r}} \frac{1}{\ln\left\{2\left(\frac{1+\sqrt{k'}}{1-\sqrt{k'}}\right)\right\}}\Omega, \quad \text{for } w/h > 0.5 \end{aligned}\right\} \tag{2.9}$$

where, $k = \text{sech}\left(\frac{\pi w}{2h}\right)$ and $k' = \tanh\left(\frac{\pi w}{2h}\right)$

2.3.1 Suspended Stripline

As the frequency is increased above the microwave range, it becomes necessary to use progressively thinner substrates to ensure that the propagation is confined to the quasi-TEM mode without the presence of any higher order mode. The width of the strip conductor in conventional planar transmission line structures becomes narrower, posing fabrication problems. To circumvent such problems, some modified version of the stripline and the microstrip have been evolved, of which, suspended stripline, suspended microstrip, and inverted microstrip are the more common ones. These structures incorporate an air gap between the dielectric substrate carrying the strip conductor and the ground plane. The presence of air gap in these structures reduces the concentration of electromagnetic energy near the ground plane. Hence, the conductor loss is far less compared with that in the microstrip. Further, the air gap reduces the effective dielectric constant of the medium, which results in an increased width of the strip conductor. The dimensional tolerances in these structures can be relaxed, resulting in considerable fabrication ease at higher microwave and even millimeter-wave frequencies.

The suspended stripline is shown in Figure 2.4. A thin dielectric substrate containing the strip conductor is suspended midway between the top and bottom ground plane, which creates an air gap on either side of the substrate. In practical version, the structure is provided with sidewalls with groove to support the substrate. The dominant mode of propagation is quasi-TEM. The introduction of air gap results in the reduction of the effective dielectric constant of the propagation medium. This configuration, therefore, permits larger circuit dimensions leading to relaxed dimensional tolerances and increased accuracy of circuit fabrication as compared with the microstrip. The presence of the air gap also reduces the conductor loss in the ground plane, because most of the electromagnetic energy that gets concentrated in the dielectric constant can be made close to that of air, thereby extending the frequency range of operation in the dominant mode, which reduces to nearly TEM.

Suspended substrate stripline technology has been used extensively in the realization of broadband filters and directional couplers for electromagnetic applications, multiplexers, antenna feeders, and other passive components both in the microwave and millimeter-wave bands up to 100GHz. Some passive components have also been reported in the 140 and 220GHz frequency bands.

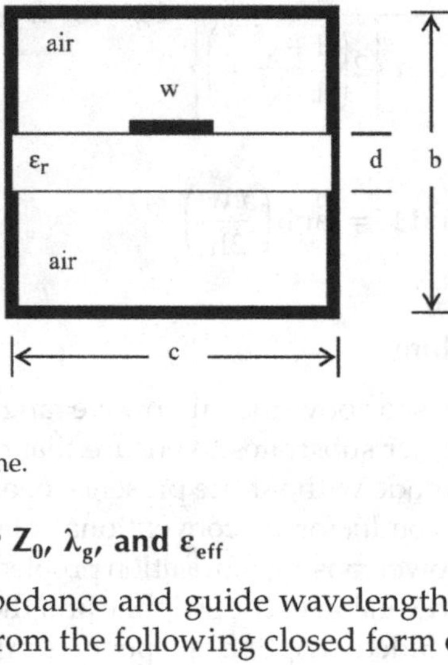

FIGURE 2.4
Shielded suspended stripline.

2.3.2 Expressions for Z_0, λ_g, and ε_{eff}

The characteristic impedance and guide wavelength for a suspended stripline can be obtained from the following closed form expressions:

For $0 < w < c/2$,

$$\lambda_g = \frac{\lambda_0}{\sqrt{\varepsilon_{eff}}} \tag{2.10}$$

where $\sqrt{\varepsilon_{eff}} = \left[1 + \{E - F\ln(w/b)\}\ln\left(1/\sqrt{\varepsilon_r}\right)\right]^{-1}$

$$Z_0 = 60\pi\left[V + R\ln\left\{\frac{6}{w/b}\right\} + \left\{1 + \frac{4}{(w/b)^2}\right\}^{1/2}\right]$$

where

$E = 0.2077 + 1.2177\,(d/b) - 0.08364\,(c/b)$

$F = 0.03451 - 0.1031\,(d/b) + 0.07142\,(c/b)$

$V = -1.7866 - 0.2035\,(d/b) - 0.4750\,(c/b)$
$R = 1.085 + 0.1007\,(d/b) - 0.09457\,(c/b)$

For $c/2 < w < c$,

$$\lambda_g = \frac{\lambda_0}{\sqrt{\varepsilon_{eff}}} \tag{2.11}$$

where $\sqrt{\varepsilon_{eff}} = \left[1 + \{E - F\ln(w/b)\}\ln\left(1/\sqrt{\varepsilon_r}\right)\right]^{-1}$

$$Z_0 = 120\pi\left[V + R\,\{(w/b) + 1.393 + 0.667\ln((w/b) + 1.444)\}^{-1}\right]$$

where
$E = 0.464 + 0.9467\,(d/b) - 0.2063\,(c/b)$
$F = 0.1424 - 0.3017\,(d/b) + 0.02411\,(c/b)$
$V = -0.6301 - 0.07082\,(d/b) + 0.247\,(c/b)$
$R = 1.9492 + 0.1553\,(d/b) - 0.5123\,(c/b)$

These expressions are valid for $1 \le c/b \le 2.5$, $1 < \varepsilon_r < 4$, 4, $0.1 < d/b < 0.5$. The accuracy of these expressions is within 2% for $0 < w < c/2$ and ±3% for $c/2 < w < c$ compared to rigorous methods.

2.3.3 Summary of Stripline

- ε_{eff} can be made close to 1 using thin substrates.
  a. Higher impedances (~200 Ω) can be realized
  b. Less dispersion
- Nearly TEM propagation even at mm-wave frequencies. Typically, Q ~ 500.
- A slight change in ε_r has second order effects on ε_{eff}.
- Dimensional tolerances required are less stringent as compared to waveguides at mm-wave frequencies.
- Circuit dimensions are compatible with beam lead and chip devices, thus offering potential for construction of both passive and active mm-wave integrated circuits.
- Permits easy transition to rectangular waveguide and other planar transmission lines.
- Provides greater flexibility in circuit design since conductors can be printed on both sides of the substrate.

- Suspended microstripline, normally used for filter applications and only very seldom for circuit applications.
- The reduced substrate thickness leads to **lower dielectric losses,** which makes this line attractive for **low pass filters**.
- Because of the small substrate height, the dispersion of this line is smaller than in the case of the conventional microstripline.

2.4 Microstripline

A cross-sectional view of a microstripline is shown in Figure 2.5. It consists of a dielectric substrate with a strip conductor on one side and the ground plane on the opposite side. Unlike the stripline, the microstrip is basically an open structure and requires high dielectric constant substrates to confine the electromagnetic fields near the strip conductor. Moreover, the microstripline is an inhomogeneous structure. Due to the composite nature of the dielectric interface, propagation cannot be true TEM, that is, a pure TEM mode cannot exist. It is not possible to satisfy the boundary conditions for this mode at the surface of the dielectric. However, at low frequencies, the mode of propagation closely resembles the TEM mode, and hence, is termed the quasi-TEM mode. The electric and magnetic field lines are concentrated predominantly in the dielectric substrate beneath the strip conductor and are somewhat less in the air region above.

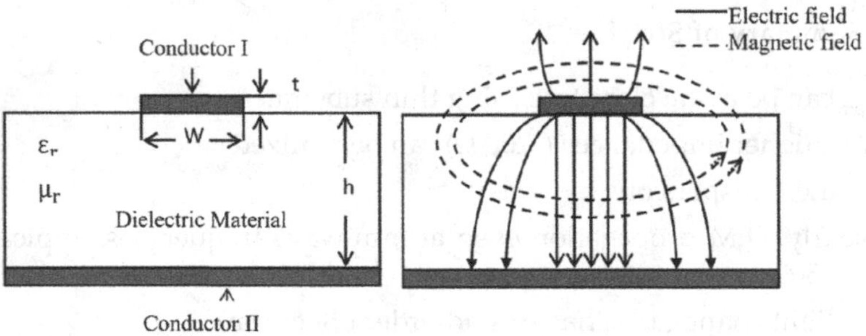

FIGURE 2.5
Cross-sectional view and field configurations of microstrip.

The larger the relative dielectric constant (ε_r) of the substrate, the greater would be the concentration of energy in the substrate region.

Being an open structure, it is easily amenable for series mounting of discrete devices and to make minor adjustments after the circuit has been fabricated. However, care has to be taken to minimize radiation loss or interference due to nearby conductors. The use of high dielectric constant substrates could be advantageous as they would reduce the phase velocity and circuit dimensions. The analysis of the microstrip is a little complicated since the structure becomes a mixed dielectric transmission line.

The microstrip is a versatile transmission line for microwave and millimeter-wave integrated circuits up to 35 GHz. Its frequency range of operation can be extended up to 94 GHz by shielding the structure and using thin dielectric constant substrates.

In a microstripline, lower dielectric constant allows higher achievable Q as well as higher operating frequency. For sufficiently small values of dielectric constant of the substrate, the maximum Q in a microstripline can be limited due to increased radiation losses. Thinner substrates permit higher frequencies to be used, but at the expense of degradation in Q-factor, especially at lower frequencies.

A major advantage of the microstrip is that its surface is accessible for mounting passive as well as active discrete devices. It is also a versatile medium for realizing a variety of circuit forms and combining several circuit functions.

Practical microstrip circuits are housed in a shielded enclosure to provide electromagnetic shielding and suppress radiation. Figure 2.6 shows the cross section of a shielded microstrip. The dimensions of the enclosure are selected such that the waveguide modes are below cutoff, and the top and sidewalls have practically no effect on the propagation characteristics.

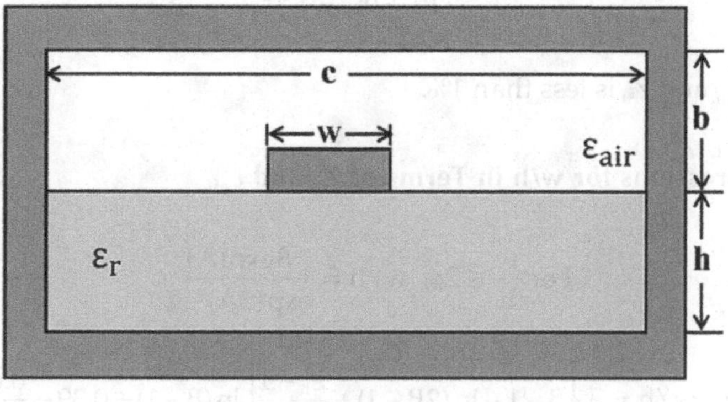

FIGURE 2.6
Shielded microstripline.

2.4.1 Applications of Microstripline

Microstripline–based filters, impedance transformers, hybrids, couplers, power dividers and combiners, delay lines, baluns, circulators, and antennas are extensively used in microwave systems, including measurement instruments where the demand on low-loss and high-power characteristics is not severe.

When the size of the microstrip section is reduced to dimensions much smaller than the wavelength, it can be used as a lumped element. Microstrip sections in lumped and distributed forms are commonly used in passive and active hybrid and monolithic integrated circuits. Examples of passive circuits include filters, impedance transformers, hybrids, couplers, power dividers and combiners, delay lines, and baluns. Amplifiers, oscillators, mixers, and control circuits employing solid-state devices are active circuits. Microstriplines have important applications in high-temperature superconducting MICs.

2.4.2 Expressions of Z_0 and ε_{eff}

$$\text{For } \frac{w}{h} \le 1, \quad Z_0 = \frac{60}{\sqrt{\varepsilon_{eff}}} \ln\left(\frac{8h}{w} + 0.25\frac{w}{h}\right) \tag{2.12}$$

$$\text{For } \frac{w}{h} \ge 1; \quad Z_0 = \frac{120\pi}{\sqrt{\varepsilon_{eff}}}\left\{\frac{w}{h} + 1.393 + 0.667\ln\left(\frac{w}{h} + 1.444\right)\right\}^{-1}$$

$$\text{where } \varepsilon_{eff} = \frac{\varepsilon_r + 1}{2} + \frac{\varepsilon_r - 1}{2} F(w/h)$$

$$F(w/h) = \begin{cases} (1 + 12h/w)^{-1/2} + 0.04(1 - w/h)^2, & \frac{w}{h} \le 1 \\ (1 + 12h/w)^{-1/2}, & \frac{w}{h} \ge 1 \end{cases} \tag{2.13}$$

Error in ε_{eff} and Z_0 is less than 1%.

2.4.3 Expressions for w/h in Terms of Z_0 and ε_{eff}

$$\text{For } \frac{w}{h} \le 2; \quad w/h = \frac{8\exp(A)}{\exp(2A) - 2} \tag{2.14}$$

$$\text{For } \frac{w}{h} \ge 2; \quad w/h = \frac{2}{\pi}\left\{B - 1 - \ln(2B - 1) + \frac{\varepsilon_r - 1}{2\varepsilon_r}\left[\ln(B - 1) + 0.39 - \frac{0.61}{\varepsilon_r}\right]\right\}$$

$$\text{where } A = \frac{Z_0}{60}\left\{\frac{\varepsilon_r + 1}{2}\right\}^{1/2} + \left(\frac{\varepsilon_r - 1}{\varepsilon_r + 1}\right)\left\{0.23 + \frac{0.11}{\varepsilon_r}\right\} \text{ and } B = \frac{60\pi^2}{Z_0\sqrt{\varepsilon_r}}$$

Error in w/h is less than 1%. These formulae are based on the assumption that the strip conductor thickness "t" is negligible.

2.4.4 Effect of Strip Thickness (t)

Formulae for Z_0 and ε_{eff} with finite strip thickness (t):

$$\text{For } \frac{w}{h} \le 1; \quad Z_0 = \frac{60}{\sqrt{\varepsilon_{eff}}} \ln\left(\frac{8h}{W_{ef}} + 0.25\frac{W_{ef}}{h}\right) \tag{2.15}$$

$$\text{For } \frac{w}{h} \ge 1; \quad Z_0 = \frac{120\pi}{\sqrt{\varepsilon_{eff}}}\left\{\frac{W_{ef}}{h} + 1.393 + 0.667\ln\left(\frac{W_{ef}}{h} + 1.444\right)\right\}^{-1}$$

$$\text{where } \frac{W_{ef}}{h} = \frac{w}{h} + \frac{1.25}{\pi}\frac{t}{h}\left(1 + \ln\frac{4\pi w}{t}\right); \quad w/h \le 1/2\pi$$

$$\frac{W_{ef}}{h} = \frac{w}{h} + \frac{1.25}{\pi}\frac{t}{h}\left(1 + \ln\frac{2h}{t}\right); \quad w/h \ge 1/2\pi \tag{2.16}$$

$$\varepsilon_{eff} = \frac{\varepsilon_r + 1}{2} + \frac{\varepsilon_r - 1}{2}F(w/h) - C \tag{2.17}$$

$$\text{where } C = \left(\frac{\varepsilon_r - 1}{4.6}\right)\frac{t/h}{\sqrt{w/h}}.$$

The effect of thickness on Z and ε_{eff} is insignificant for small values of t/h ($t/h \le 0.005$, $2 \le \varepsilon_r \le 10$, and $w/h \ge 0.1$). However, the effect of strip thickness is significant on conductor loss in the microstripline.

2.4.5 Losses

Considering microstrip as a quasi-TEM line, the attenuation due to dielectric loss can be determined as

$$\alpha_d = \frac{\kappa_0 \varepsilon_r (\varepsilon_{eff} - 1)\tan\delta}{2\sqrt{\varepsilon_{eff}}\,(\varepsilon_{eff} - 1)} \text{Np/m} \tag{2.18}$$

where $\tan\delta$ is the loss tangent of the dielectric.

The attenuation due to **conductor loss** is given approximately by

$$\alpha_c = \frac{R_s}{zw} \text{Np/m} \tag{2.19}$$

where $R_s = \sqrt{\omega\mu_0/2\sigma}$ is the surface resistivity of the conductor. For most dielectric substrates, conductor loss is much more significant than the dielectric loss.

2.4.6 Power-Handling Capability

Microstrip transmission lines can be used for several medium-power applications. A 50 Ω microstrip on a 25 mil substrate (alumina) can handle a few kW of power.

The power-handling capability of a microstrip, like that of any other dielectric-filled transmission line, is limited by the heating caused because of **ohmic and dielectric losses** and by **the dielectric breakdown**. The increase in temperature due to conductor and dielectric losses limits the **average power** of the microstripline, while the breakdown between the strip conductor and the ground plane limits the **peak power**.

2.4.7 Average Power

The average power-handling capability of a microstrip is determined by the temperature rise of the strip conductor and the supporting substrate. The parameters that play major roles in the calculation of average power capability are:

1. Transmission line losses
2. Thermal conductivity of the substrate material
3. Surface area of the strip conductor
4. Ambient temperature, that is, the temperature of the medium surrounding the microstrip.

Therefore, dielectric substrates with a low loss tangent and large conductivity will power the capability of microstriplines.

2.4.8 Density of Heat Flow due to Conductor Loss

Loss of electromagnetic power in the strip conductor generates in the strip. Because of the good heat conductivity of the strip metal, heat generation is uniform along the width of the conductor. Since the ground plane is held at ambient temperature, this heat flows from the strip conductor to the ground plane through the substrate. The heat flow can be calculated by considering analogous electric field distribution.

The heat flow in the microstrip structure corresponds to the electrostatic of the microstrip as shown in Figure 2.7.

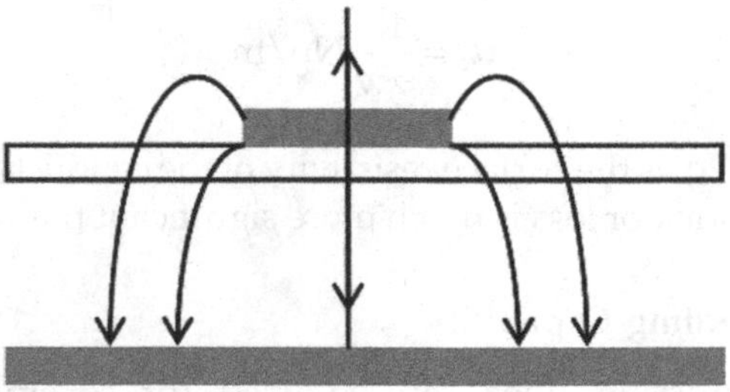

FIGURE 2.7
Heat flow in microstrip.

The electric field (the thermal field in the case of heat flow) spreads near the ground plane. To account for the increase in area normal flow lines, we adopt a parallel plate model, which is shown in Figure 2.8 (Table 2.1).

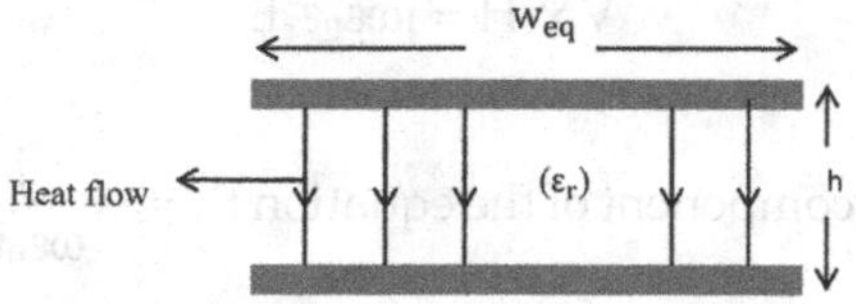

FIGURE 2.8
Electric field (the thermal field in the case of heat flow).

TABLE 2.1
Analogy between Heat Flow Field and Electric Field

S. No.	Heat Flow Field	Electric Field
1.	Temperature T (°C)	Potential V (V)
2.	Temperature gradient T_g (°C/m)	Electric field E (v/m)
3.	Heat flow rate Q (W)	Flux Φ (C)
4.	Density of heat flow q (W/m²)	Flux density D (C/m²)
5.	Thermal conductivity K (W/m °C)	Permittivity ε (C/m °C)
6.	Density of heat generated	Change density
7.	$q = -k \nabla T$	$D = -\varepsilon \nabla v$
8.	$\nabla q = \rho h$	$\nabla \vec{D} = \rho$

2.4.9 Microstrip–Quasi-TEM Mode

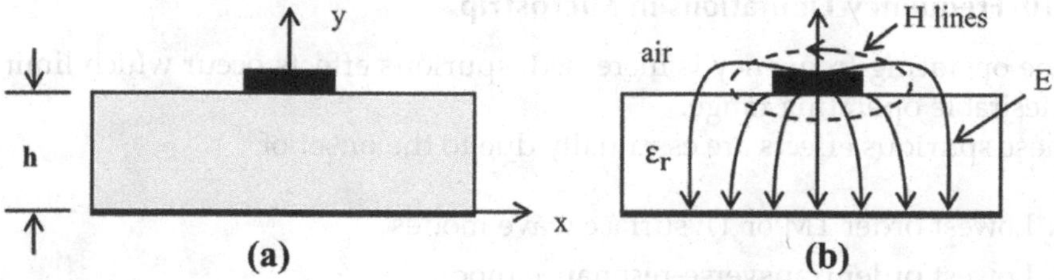

FIGURE 2.9
(a) Microstrip and (b) E-field and H-field configurations.

The microstripline and field configurations are shown in Figure 2.9a,b, respectively. Due to the mismatch in the velocity of propagation at the interface between air and dielectric, pure TEM mode cannot exist in a microstrip.

- Consider Maxwell's Curl H equation in a source-free dielectric medium:
  Time dependence of $e^{i\omega t}$

$$\nabla \times \vec{H} = j\omega\varepsilon_0\varepsilon_r\vec{E} \tag{2.20}$$

Consider the x-component of the equation $E_x = -\frac{j}{\omega\varepsilon_0\varepsilon_r}\hat{x}\left(\frac{\partial H_z}{\partial y} - \frac{\partial H_y}{\partial z}\right)$

At $y = h$, E_x must be continuous.

$$\frac{1}{\varepsilon_r}\left(\frac{\partial H_z}{\partial y} - \frac{\partial H_y}{\partial z}\right)\Bigg|_{y=h^-} = \left(\frac{\partial H_z}{\partial y} - \frac{\partial H_y}{\partial z}\right)\Bigg|_{y=h^+} \tag{2.21}$$

At $y = h$, the normal component B or H_y must be continuous.
Use $H_y\big|_{y=h^+} = H_y\big|_{y=h^-} = H_y$ in (2.21)

$$\varepsilon_r\frac{\partial H_z}{\partial y}\Bigg|_{y=h^+} - \frac{\partial H_z}{\partial y}\Bigg|_{y=h^-} = (\varepsilon_r - 1)\frac{\partial H_y}{\partial z} \tag{2.22}$$

Since $H_y \neq 0$, it is necessary that $H_z \neq 0$. Similarly, it can be shown that $E_z \neq 0$.
Microstripline does not support pure TEM mode.
However, E_z and H_z components $\lll$ transverse field components.
Hence, the dominant mode is called the quasi-TEM mode.

2.4.10 Frequency Limitations in Microstrip

As the operating frequency is increased, spurious effects occur which limit the desirable operating range.

These spurious effects are essentially due to the onset of

1. Lowest order TM or TE surface wave modes
2. Lowest order transverse-resonance mode

- TM-mode limitation ($w/h \ll 1$)
  i. Surface waves exist on a microstrip with a very narrow strip conductor.
  ii. These waves couple to the quasi-TEM mode when their phase velocities are close to each other. The frequency at which this occurs is given by

$$f_{c\,TM} = \frac{C\tan^{-1}(\varepsilon_r)}{\sqrt{2}\,\pi h\,\sqrt{\varepsilon_r - 1}} \tag{2.23}$$

- TE-mode limitation (w/h ≪ 1)
  The lowest order TE mode has a cutoff frequency

$$f_{c\,TE} = \frac{C}{4h\,\sqrt{\varepsilon_r - 1}} \tag{2.24}$$

2.4.11 Microstrip Design Summary

- Select the substrate such that the TM-mode effect is avoided.
  The maximum intended frequency (operating frequency) should be less than

$$f_{TEM1} = \frac{C\tan^{-1}(\varepsilon_r)}{\sqrt{2}\;\pi h\,\sqrt{\varepsilon_r - 1}} \tag{2.25}$$

  Also ensure that the parasitic coupling is sufficiently low.
- Carry out a "first cut" design on the basis of static TEM and this yields "W" and ε_{eff}.
- Check that the first order transverse resonance cannot be excited at the highest frequency to be used.

$$f_{CT} = \frac{c}{\sqrt{\varepsilon_r}\,(2w + 0.8h)} \tag{2.26}$$

  In the event that such a resonance might occur, either redesign (using narrower microstriplines) or, if possible, incorporate slots in the metal strips.
- Calculate total losses and the total Q-factor, and check that it meets the circuit requirements.

2.5 Suspended Microstripline and Inverted Microstripline

The suspended microstrip is a variant of the microstrip with an air gap between the substrate and the ground plane, which is shown in Figure 2.10. The inverted microstrip differs from the suspended microstrip in which the strip conductor is situated on a lower surface of the dielectric

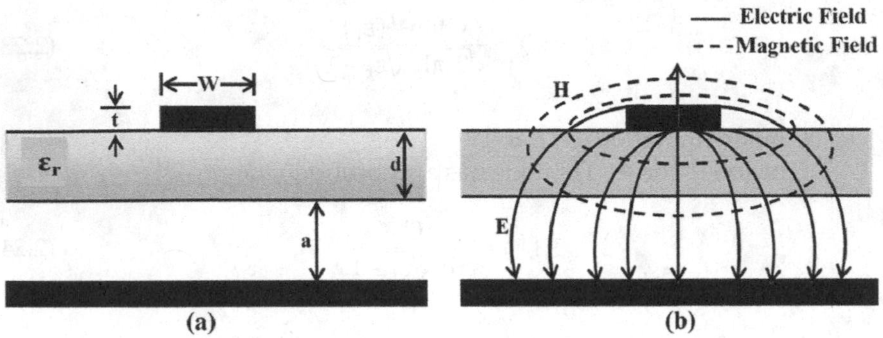

FIGURE 2.10
(a) Suspended microstripline and (b) electric and magnetic field distributions.

substrate facing the ground plane, which is shown in Figure 2.11. Both suspended and inverted microstriplines can also be viewed as special cases of the suspended stripline, with one of the ground planes removed. These structures retain the advantages of the suspended stripline in terms of achieving larger strip dimensions and lower dissipative losses with respect to the microstrip. In comparison with the suspended microstrip, the inverted microstrip has the advantages of reduced radiation loss by virtue of having the strip conductor below the substrate. However, because the air gap involved is too small (typically on the order of 1 mm or less), incorporation of semiconductor devices becomes very difficult in this configuration.

The suspended and inverted microstriplines are used for precision low-loss lines. They allow a wider range of achievable impedances. In addition, the air–dielectric interface is essentially lossless and dominates the dispersion characteristics.

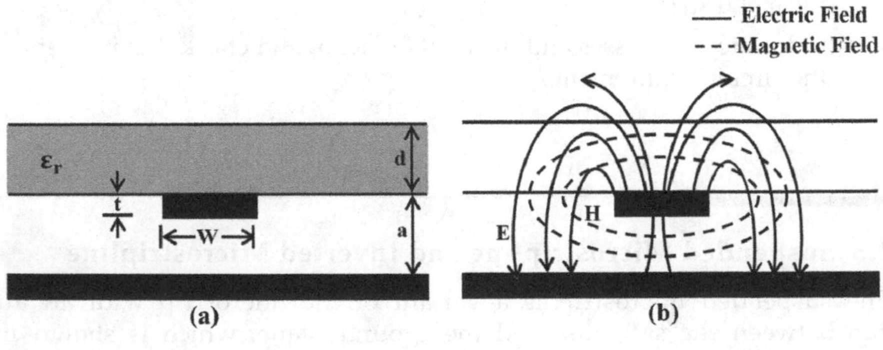

FIGURE 2.11
(a) Inverted microstripline and (b) electric and magnetic field distributions.

2.5.1 Characteristic Impedance Z_0 and Effective Dielectric Constant ε_{eff} For Microstrip,

$$Z_0 = \left(60/\sqrt{\varepsilon_{eff}}\right)\ln\left[f(u)/u + \left(1+4/u^2\right)^{1/2}\right] \tag{2.27}$$

where

$$f(u) = 6 + (2\pi - 6)\exp\left[-(30.666/u)^{0.7528}\right]$$

$$u = w/(a+d)$$

$$\sqrt{\varepsilon_{eff}} = \left[1 + (d/a)(a_1 - b_1\ln(w/a))\left(1/\sqrt{\varepsilon_r} - 1\right)\right]^{-1} \tag{2.28}$$

where

$$a_1 = [0.8621 - 0.1251\ln(d/a)]^4$$

$$b_1 = [0.4986 - 0.1397\ln(d/a)]^4$$

For Inverted Microstrip,

The effective dielectric constant of inverted microstrip is expressed by

$$\sqrt{\varepsilon_{eff} = \left[1 + (d/a)(a_2 - b_2\ln(w/a))\left(1/\sqrt{\varepsilon_r} - 1\right)\right]^{-2}} \tag{2.29}$$

where

$$a_1 = [0.5173 - 0.1515\ln(d/a)]^2$$

$$b_1 = [0.3092 - 0.1047\ln(d/a)]^2$$

The accuracy is reported to be within ±1% for $1 < w/a \le 8$, $0.2 < d/a \le 1$, and $\varepsilon_r \le 6$.

2.6 Slotline

Slotlines are open structures (realized usually on high dielectric constant substrates). Finlines are slot lines that are realized on low dielectric constant substrates enclosed in a standard rectangular waveguide.

The slotline structure was first proposed by Cohn in 1968 and an extensive account has been given by K.C. Gupta et al. The cross section of the *slotline* is shown in Figure 2.12a. It consists of a slot etched from the conducting layer on one side of the dielectric substrate and the other side being bare.

The electric field is essentially transverse to the slot because of the voltage difference across the gap. The magnetic field in a slotline has both a longitudinal and a transverse component. This implies that a rotating magnetic field vector is present at some point, which can be used for coupling to the gyromagnetic effects in ferrites. The dominant mode of propagation is essentially non-TEM, and hence, the characteristic impedance and guide wavelength are dependent on frequency, although at a slow rate. This feature is in contrast with the TEM nature of the stripline whose propagation parameters are independent of frequency, and also with the quasi-TEM nature of the microstripline whose propagation parameters are, to a first order, independent of frequency. The slotline differs also from the waveguide in that it has no cutoff frequency.

Structurally, the slotline is complementary to the microstrip. Compared with the microstrip, the slot line is much more dispersive. Slotline is not a TEM structure. However, its single-sided planar nature makes it of interest for modern fabrication techniques. This structure requires a high dielectric constant material ($\varepsilon_r \geq 10$) to confine the fields and minimize radiation.

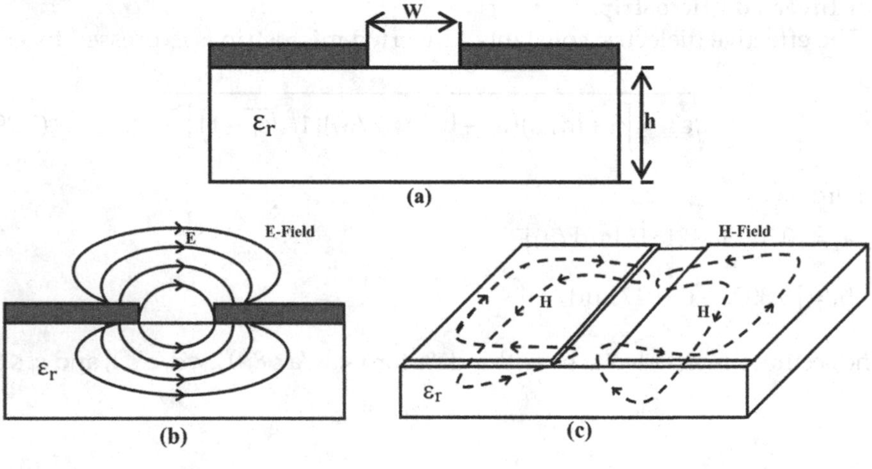

FIGURE 2.12
(a) Cross section of slotline, (b) electric field configuration, and (c) magnetic field configuration.

2.6.1 Advantages

- In a slotline, both the conductors are in one plane and hence, shunt mounting of discrete devices (active and passive) across the line is very conventional.
- Higher impedance can be realized.
- Slotlines are useful for the design of several ferrite components.

- Short-circuited slotlines are used for inductances and open-circuited slotlines for capacitance.

2.6.2 Disadvantages

- Coaxial-to-microstrip transitions are relatively easier to fabricate than coaxial-to-slotline transitions.
- Q-factor is around 100, which is significantly lower than the other structures.
- Characteristic impedances below about 60 Ω are difficult to realize.
- More dispersive than microstrip.
- Circuit structures often involve difficult registration problems (particularly with metallization on the opposite side to the slot).

2.7 Comparison between Slot Line and Microstrip Line

- Due to the presence of Hx and Hy components, a region of elliptically polarized magnetic field exists near the slot. Useful for the fabrication of nonreciprocal ferrite circuits.
- Since electric field lines run across the slot, shunt mounting of discrete components (active as well as passive) across the line is made easy.
- For a slot line, Z_0 increases with the width of the slot, whereas in the case of a microstrip, Z_0 decreases with the width of the strip. Thus, high impedance lines, which require a very thin line to get photo etched in a microstrip configuration, are more easily achieved with the slot line (Table 2.2).

TABLE 2.2

Comparison between Slotline and Microstrip

S. No.	Slot Line	Microstrip
1.	Essentially a non-TEM transmission line and therefore Z_0 vary considerably with frequency.	Comparatively, the microstrip Z_0 exhibits only a small variation with frequency.
2.	Larger fraction of the power flow in the air region than that in the microstripline. That is, for the same dielectric substrate, the effective dielectric constant is lower than that for the microstripline.	

2.8 Coplanar Waveguide (CPW)

Coplanar waveguide (CPW), proposed by C.P. Wen in 1969, is another basic transmission line where all the conductors are situated on one side of the dielectric substrate. It consists of a strip conductor separated by a slot on either side from the two adjacent ground planes as shown in Figure 2.13a. Its electrical and magnetic field distributions are shown in Figure 2.13b.

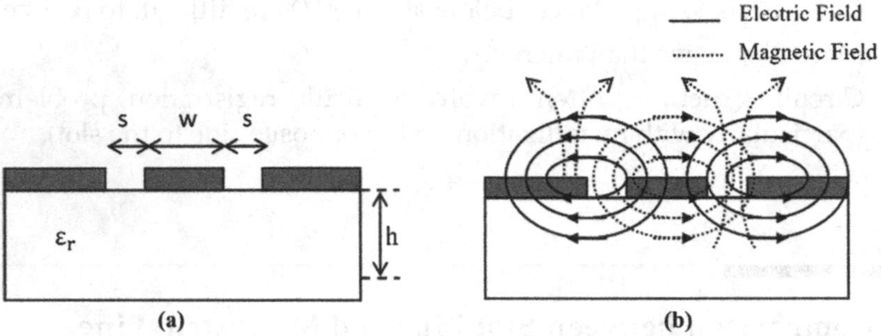

FIGURE 2.13
(a) Cross-sectional view of CPW and (b) field configurations.

This is a guide medium considered as an alternative to the microstripline in millimeter-wave integrated circuits. Its principal advantage is the location of the signal grounds on the same substrate surface as the signal lines. This eliminates the need for via holes and thus simplifies the fabrication process. It also permits easy connection of both series and shunt components. Since the structure dimensions are almost independent of substrate thickness, the dimensions are chosen very small. It has small dispersion due to the fact that the field is mainly concentrated in the spacing between the conductors and does not change much with increasing frequency. These advantages make CPW an attractive choice for the development of MMICs. For an MMIC design, the effectiveness of modern computer-aided design (CAD) methods relies heavily on the availability of accurate analysis and synthesis models.

Here, the dominant mode of propagation resembles quasi-TEM. However, at higher frequencies, the field becomes non-TEM because of the contribution due to the longitudinal component of the magnetic field. The magnetic field in the slots at the air–dielectric interface is elliptically polarized and hence the CPW becomes suitable for nonreciprocal ferrite devices.

The CPW structure is proposed to realize the detectors, balance mixers, phase shifters, delay lines, filters, and directional couplers. Impedances in the range of 40–150 Ω are realizable in CPW. Its Q-factor is low of the order of 100 as compared with a value of about 250 in the microstrip.

2.8.1 Calculation of Phase Velocity (ν_p) and Z_0 for CPW with Infinitely Thick Substrate

The phase velocity (ν_p) of a conventional CPW is

$$v_p = \frac{v_0}{\sqrt{\varepsilon_{eff}}} \tag{2.30}$$

where the effective permittivity $\varepsilon_{eff} = \frac{C}{C_a} = \frac{\varepsilon_r + 1}{2}$. C is the capacitance per unit length of the structure, C_a is the capacitance per unit length of the length of the same structure with all dielectrics replaced by air, and ν_0 is the velocity of light. Hence,

$$v_p = v_0 \sqrt{\frac{2}{\varepsilon_r + 1}} \tag{2.31}$$

The characteristic impedance of CPW is obtained as

$$Z_0 = \frac{30\pi}{\sqrt{\varepsilon_{eff}}} \frac{K'(k_1)}{K(k_1)} \Omega \tag{2.32}$$

where

$$\frac{k(k)}{k'(k)} = \frac{\pi}{\ln\left[2\left(1+\sqrt{k'}\right)/\left(1-\sqrt{k'}\right)\right]} \quad \text{for } 0 \le k \le 0.707$$

$$\frac{k(k)}{k'(k)} = \frac{1}{\pi}\ln\left[2\left(1+\sqrt{k}\right)/\left(1-\sqrt{k}\right)\right] \quad \text{for } 0.707 \le k \le 1$$

Here, the ratio K/K′ varies from 0 to ∞ as k varies from 0 to 1.

2.8.2 Advantages

- Uniplanar configuration.
- Ease of shunt and series connections without having a necessity for drilling.
- Easier fabrication.
- Good grounding for active devices.
- Low radiation.

- Low dispersion and smaller losses (the field is mainly concentrated in spacing between conductors and does not change much with increasing frequency).
- Reduced component dimensions (Miniaturization).
- Avoidance of the need for thin fragile substrates (in MMIC applications, the semiconductor substrates are usually thin and fragile).
- Structure dimensions are almost independent of substrate thickness.
- CPW is an attractive choice for MMIC developments.
- Dominant mode resembles the quasi-TEM mode. At higher frequency, the field becomes non-TEM because the contribution due to the longitudinal component of the magnetic field exists.
- H-field in the slots at air–dielectric interface is elliptically polarized. Hence, suitable for ferrite.

2.8.3 Disadvantages

- Coplanar lines suffer from disadvantages of larger size, parasitic odd mode, lower power-handling capability, poor heat transfer for active devices, and field non-confinement. For improving both the mechanical strength and the power-handling capability, a conductor-backed CPW can be used. Moreover, conductor-backed CPW and those with finite width ground planes are found to be less dispersive than others.

2.8.4 Applications

As MESFETs (Metal Semiconductor Field Effect Transistor) are coplanar in nature, they can easily be connected to coplanar lines. Coaxial-to-CPW and microstrip-to-CPW transitions can be easily achieved. A CPW can propagate a signal in two modes: (i) an unbalanced signal in the even mode of coupled slot lines and (ii) a balanced signal in the odd mode of coupled slot lines. The impedances for these modes (balanced and unbalanced) are different. The slotline-to-CPW junction excites a balanced signal in CPW. But when a CPW is fed from a coaxial line or through a microstrip, an unbalanced signal is launched on the CPW. Hence, a microstrip-CPW-slotline or a coax-CPW-slotline combination may be used in circuits where both the balanced and the unbalanced signals are employed.

Both balanced and unbalanced modes are possible in CPW. Thus, CPW finds extensive applications in balanced mixers, double-balanced mixers, balanced modulators, and balanced frequency multipliers where both the balanced and unbalanced signals are present. The magic-T can be realized in a CPW–slotline configuration.

2.9 Coplanar Strips

The coplanar strips (CPS) structure, as shown in Figure 2.14a, is similar to a paired wire transmission line structure and is the complementary structure to CPW. It consists of two parallel coupled-strip conductors located on the same surface of the dielectric substrates, with one of them serving as the ground plane. The electric and magnetic field configurations are shown in Figure 2.14b. As in the CPW, the dominant mode is a form of the quasi-travelling wave oscillator (TWO) mode. This configuration is suitable for microwave monolithic integrated circuits built on semiconducting substrates.

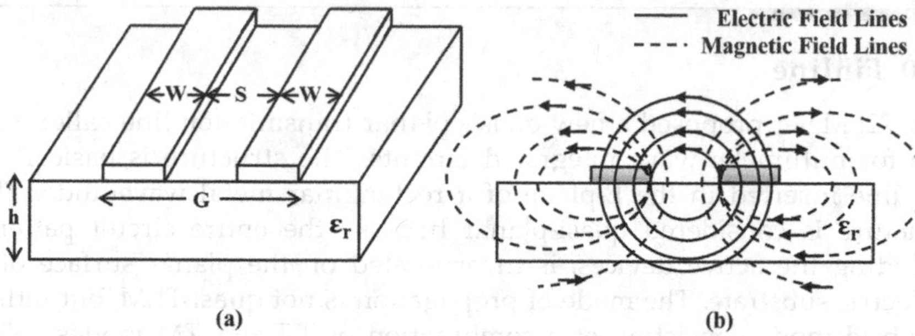

FIGURE 2.14
(a) CPS structure and (b) electric and magnetic distributions in CPS.

Other transmission structures used in MMICs are the microstrip and the CPW. Compared with the microstrip, CPS offers much higher impedance levels, typically in the range of 40–250 Ω. The CPS permits mounting active devices in series as well as in shunt configurations, whereas the microstrip and its variants are convenient only for series mounting. However, the loss in this structure is higher than that in a microstrip for the same characteristic impedance. In the lower range of microwave frequencies, the CPS is also useful for carrying signals for high-speed computer circuits.

2.9.1 Calculation of Z_0 and ε_{eff}

The characteristic impedance Z_0 and effective dielectric constant ε_{eff} of the coplanar strips are

$$Z_0 = \frac{120\pi}{\sqrt{\varepsilon_{eff}}} \frac{K(k)}{K(k')} \tag{2.33}$$

$$\varepsilon_{eff} = 1 + \frac{\varepsilon_r - 1}{2} \frac{K(k')}{K(k)} \frac{K(k_1)}{K(k_1')} \tag{2.34}$$

where

$$K = \frac{S}{G}, k' = \sqrt{1-k^2}, k_1' = \sqrt{1-k_1{}^2}$$

$$k_1 = \frac{\sin\left(\frac{\pi}{4}\frac{S}{h}\right)}{\sin\left(\frac{\pi}{4}\frac{G}{h}\right)}$$

2.10 Finline

In 1972, Meier proposed a new quasi- planar transmission line called fine line for millimeter-wave integrated circuits. The structure is basically a slot line inserted in the E-plane of a rectangular metal waveguide. The structure is considered quasi-planar because the entire circuit pattern, including the active devices, is incorporated on the planar surface of a dielectric substrate. The mode of propagation is not quasi-TEM, but rather a hybrid mode consisting of a combination of TE and TM modes. With a dominating TE term, the mode is designated an HE-mode, and with a dominating TM term, the mode is designated an EH-mode. If the walls are assumed to be perfectly conducting, then at cutoff, an HE-mode becomes purely TE and an EH-mode becomes purely TM. The single-mode operating bandwidth in a finline is greater than the bandwidth of the corresponding hollow waveguide.

The finline, as shown in Figure 2.15, is basically a slotline mounted in the E-plane of a standard rectangular waveguide, or we can say, the finline is a shielded slotline. The fins are usually metal foils on a thin dielectric substrate. Finline is not a TEM structure. For the dominant mode, the current flows in the axial direction. Hence, good electrical contact between the fins and the waveguide is not essential. The mode of propagation in a finline is hybrid mode. When excited by standard rectangular waveguide operating in the TE_{10} mode, this hybrid has a dominant TE term and is called 'HE-mode.'

The finline is suitable for use in microwave circuits that incorporate two-terminal devices such as diodes. As transistors are three-terminal devices, they cannot be connected to a finline. The cutoff wavelength for a finline may be determined by using the transverse-resonance method. The finline is a well-proven quasi-planar transmission line for millimeter-wave integrated circuit components in the frequency range from 30 to about 120 GHz. Due to its planar nature, it is very useful for fabricating circuits. It is also easy to interface to waveguide because of its fully shielded structure.

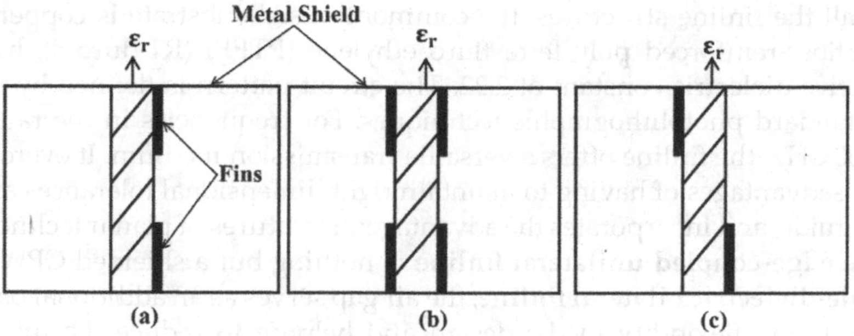

FIGURE 2.15
Basic finline structures: (a) Unilateral, (b) Bilateral, and (c) Antipodal.

Figure 2.15 shows the cross-sectional views of three basic finlines, namely, (i) Unilateral, (ii) Bilateral, and (iii) Antipodal. The bilateral finline can support two orthogonal modes, namely, the even mode and odd mode. For all practical purposes, the even mode is important because it gets excited by the incident TE_{10} mode of a hollow rectangular waveguide. Antipodal finline provides a wider range of impedance than the other finline structures. Metallization can be on a single side of the dielectric, both sides, or diagonal. The finline may be captivated in the shield and grounded or isolated.

The characteristic impedance (Z_0) of the finline can be determined by

$$Z_0 = \frac{V^2}{2P_{av}} \tag{2.35}$$

where P_{av} is the time-averaged power transported in the finline and V is the voltage across the slot which can be obtained from the slot field series expansion across the slot.

2.10.1 Basic Finline Structures

- Unilateral
- Bilateral
- Insulated
- Antipodal.

In a **unilateral finline**, the fins are located on one side of the dielectric substrate, and in a **bilateral finline**, the fins are located on both sides. The **insulated finline** incorporates a dielectric gasket to isolate the fin from the waveguide, which thus facilitates dc bias to be applied to active components. **Antipodal finline** incorporates one fin asymmetrically located on either side of the substrate. This configuration is used for realizing lower impedance values because the fins can be overlapped.

In all the finline structures, the commonly used substrate is copper-clad microfiber-reinforced poly tetra fluro ethylene (PTFE) (RT-duroid), having a relative dielectric constant of 2.22. The circuit pattern is defined by using the standard photolithographic techniques. For frequencies in the range of 30–100 GHz, the finline offers a versatile transmission medium. It overcomes the disadvantages of having to maintain tight dimensional tolerances as in a waveguide, and incorporates the advantageous features of planar technology.

The **edge-coupled unilateral finline** is nothing but a shielded CPW. In a **double-dielectric bilateral finline**, the air gap serves as an additional parameter offering flexibility in the design and helping to reduce the insertion loss. Such coupled structures, including the bilateral finline and the edge-coupled bilateral finline, are useful in the design of a number of components, especially filters.

In addition to finlines, the microstripline and the suspended stripline can be configured as E-plane transmission lines by mounting the substrate that carries the conductor pattern in the E-plane of a waveguide.

The loss in a finline (expressed in dB/λ) is reported to be nearly three times lower than that in a microstrip using the same dielectric material and waveguide mount. The suspended stripline as an E-plane transmission line has a larger ground plane spacing than a conventional line operating in the quasi-TEM mode.

The determination of propagation parameters for this line requires the use of hybrid-mode analysis (Figure 2.16).

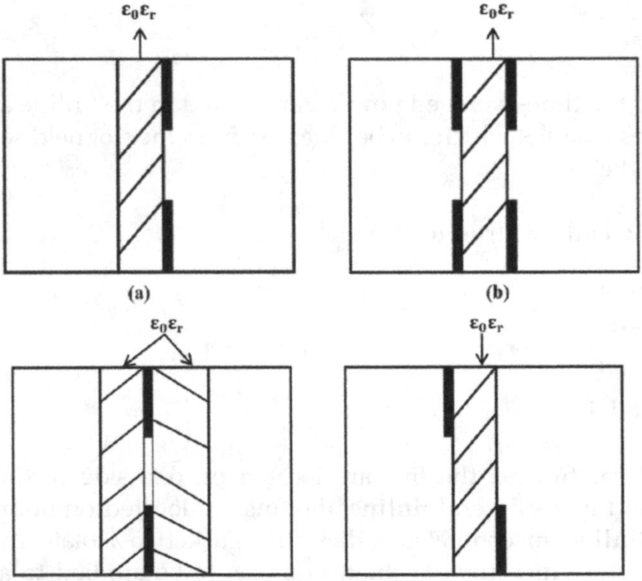

FIGURE 2.16
Finline structures: (a) Unilateral, (b) Bilateral, (c) Insulated, and (d) Antipodal.

2.10.2 Some Coupled Finline Structures

See Figure 2.17.

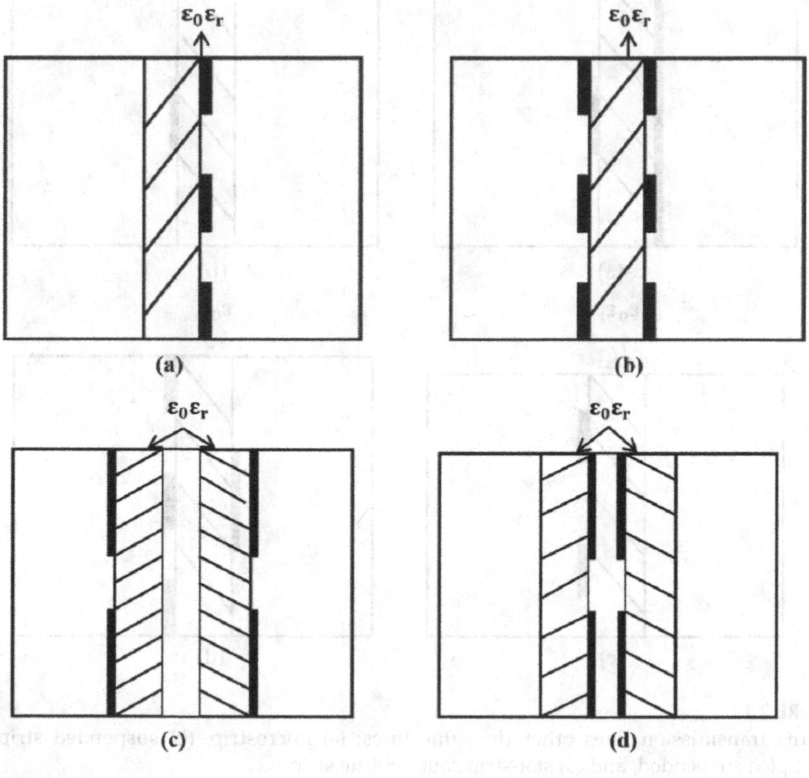

FIGURE 2.17
Some coupled finline structures: (a) Edge-coupled unilateral finline, (b) edge-coupled bilateral finline, (c) double-dielectric bilateral finline (fins facing side walls), and (d) double-dielectric bilateral finline (fins facing each other).

2.10.3 E-Plane Transmission Lines Other than Finlines

See Figure 2.18.

2.10.4 Advantages

i. The dimensions of the circuit in the 30–100 GHz frequency band are compatible with beam lead and chip devices, thus offering the potential for construction of passive and active integrated circuits.

ii. Simpler for fabrication, that is, the guide wavelength is longer than in a microstrip, thus permitting less stringent dimensional tolerances.

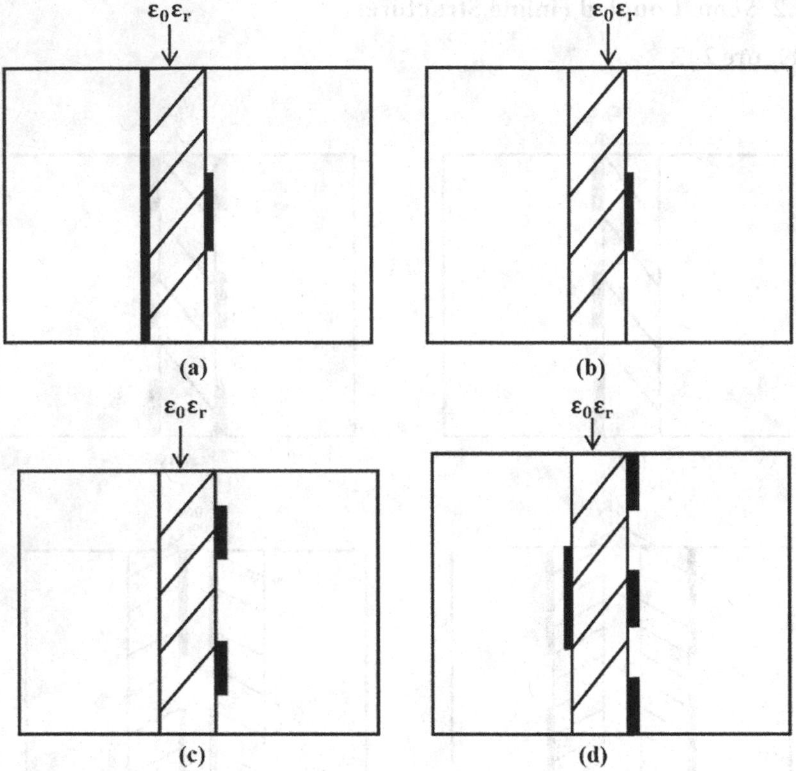

FIGURE 2.18
E-plane transmission lines other than fine lines: (a) microstrip, (b) suspended stripline, (c) coupled suspended, and (d) slot-strip coupled line stripline.

iii. It permits easy transition to a standard rectangular waveguide and operates over the entire bandwidth of the waveguide.

iv. Low-loss propagation (loss expressed in dB/λ is nearly three times lower than in microstrip using the same dielectric and waveguide mount).

v. The characteristic impedance range is remarkably large in the range of 10–400 Ω. This is more than adequate for nearly all conceivable filter or matching network applications.

2.10.5 Disadvantages

Beyond 100 GHz, finlines pose fabrication difficulties because of increasingly stringent dimensional tolerances. Dielectric integrated guides offer an alternative approach with the potential for lower losses and relaxed tolerances (Tables 2.3 and 2.4).

TABLE 2.3

Comparison of Various Planar Transmission Lines at Microwave Features

Feature	Stripline	Microstrip	CPW
Radiation emission	Low	Fair	Low
Quality factor (Q)	Fairly high	Fair	Fair
Dispersion	Low	Fairly low	Low
Thermal conductivity	Low	Good	Low
Series connections	Easy	Easy	Easy
Shunt connections	Difficult need to drill holes	Difficult need to drill holes	Easy no matching
Transition to other media	Complex machining of motherboard	Complex machining of motherboard	Easy surface mount for motherboard
Low parasitic lumped elements	Poor	Poor	Very good
Fabrication of directional couplers	Very good	Fair	Fair
Fabrication of filters	Good	Good	Fairly good
Fabrication of amplifiers	Practically impossible	Good	Good
Fabrication of isolators	Good	Good	Very difficult
Incorporation of MMICs	Poor	Good	Good

TABLE 2.4

Comparison of Various Planar Transmission Lines at Microwave Frequencies

Configurations	Typical Impedance Range (Ω)	Unloaded Q-Factor (Q_u)	Radiation loss	Dispersion	Other Features
Stripline	20–120	High (≈500)	Nil	Negligible	Excellent for passive components, inconvenient for the incorporation of chip devices
Microstripline	20–100	Medium (≈250)	Low	Small	Widely used in MICs and MMICs
Slotline	50–200	Low (≈100)	High	Large	Suitable for shunt mounting of chip devices and fabrication of nonreciprocal ferrite components.
Suspended stripline	40–150	High	Nil	Small	Suitable for high Q passive components; operation can be extended to millimeter-wave frequencies.

(*Continued*)

TABLE 2.4 (*Continued*)

Comparison of Various Planar Transmission Lines at Microwave Frequencies

Configurations	Typical Impedance Range (Ω)	Unloaded Q-Factor (Q_u)	Radiation loss	Dispersion	Other Features
Suspended microstripline	40–150	High	Low	Small	Useful at higher microwave and millimeter-wave frequencies.
Inverted microstrip	25–130	High	Low	Small	Suitable for high Q passive components, inconvenient for mounting active devices.
Coplanar Waveguide	40–150	Low	Medium	Medium	Easy connection of series and shunt elements, useful for MMICs and nonreciprocal ferrite components
Coplanar strips	40–250	Low	Medium	Medium	Easy connection of series and shunt elements, useful for MMICs.

2.11 Microwave Integrated Circuit

The use of planar circuit architecture in the forms of MICs, MMICs, and RF MEMS has opened up new opportunities in terms of reduction in size, weight, power consumption of components, as well as extension of their operating frequencies even up to the millimeter-wave band. Advances are taking place in making use of the compatibility of micro-machining technology with MICs, and planar antennas in realizing high-performance microwave and millimeter-wave products. In keeping with the advances in technology, the design approach is also undergoing rapid changes through improved digital signal processing techniques and CAD tools.

Miniaturization and the move toward higher frequencies for wireless applications are also the trends influencing the direction of RF/microwave components and system development. The expansion and merging of RF technology with MIC techniques have widened the scope for high-performance wireless products for communication. A host of wireless applications join the many well-established uses of RF/microwaves in the approximate frequency range from 500 MHz to about 100 GHz, including terrestrial and satellite-based communications, radar, heating, medical therapy/diagnostics, etc.

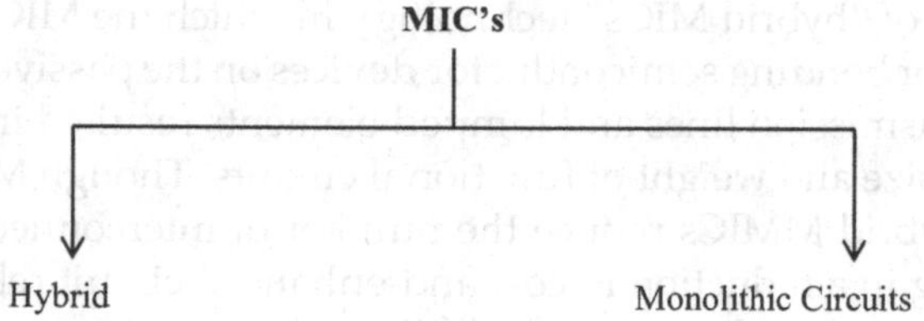

2.11.1 Hybrid MIC

- First developed in the 1960s
- Very flexible and cost-effective implementation
- One layer of metallization (TLs). Discrete components (R, L, C, transistors, etc.) are mounted on the substrate.

2.11.2 Monolithic Circuits

- New development
- The substrate is a semiconductor material. Several layers of metal dielectric and films are used.
- Active and passive components are grown on the substrate.

A typical hybrid MIC has all the transmission lines deposited on the dielectric surfaces except solid-state devices such as transmission and their passive component such as capacitors. These solid-state devices and passive elements are discrete components and are connected to the transmission line by bonding, soldering, and conducting epoxy.

The substrate of a hybrid MIC is generally low-loss insulation, used solely for supporting the circuit components and delivering the signals.

2.11.3 Technology of Hybrid MICs

The integration of miniature solid-state microwave devices with planar transmission lines has led to the evolution of MICs. Recently, a new generation of microwave components having a much higher degree of miniaturization, wider bandwidth capability, multifunction integration, and higher system reliability is possible through "monolithic technology." A monolithic millimeter wave integrated circuits (MMIC) is a microwave circuit in which the active and passive components are fabricated on the same semiconductor substrate. Monolithic circuits operating in the millimeter-wave region from 30 to 300 GHz are called monolithic millimeter-wave integrated circuits (MMICs). The additional term "monolithic" is necessary to distinguish them from the established MIC, which is a hybrid microwave circuit in which a number of discrete active and passive components are integrated on to a common substrate using solder or conductive epoxy.

The advent of "hybrid MICs" technology in which the MICs are fabricated by soldering or bonding semiconductor devices on the passive circuitry made of planar transmission lines and lumped elements resulted in a considerable reduction in size and weight of functional circuits. Though MMICs are more expensive, hybrid MMICs reduce the number of interconnections considerably, resulting in a reduction in cost and enhanced circuit reliability. Hybrid MICs offer considerable device flexibility allowing the use of a variety of planar transmission line configurations. Monolithic MICs are restricted to microstrip and coplanar line configurations.

2.11.4 Transmission Media for MICs

The selection of the substrate material in the design of any MIC depends on the type of fabrication technology, that is, hybrid or monolithic, the type of transmission line used, and the frequency band of operation.

Hybrid MICs use dielectric substrates, which are isotropic or anisotropic. The dielectric substrates of hybrid MICs must have low dissipation factor, minimum variation of dielectric constant, and dissipation factor with temperature, high thermal conductivity, polished surfaces, low loss, flexibility, and easy machinability and uniformity of thickness and dimensional stability. The important isotopic substrates used in hybrid MICs are RT-duroid ($\varepsilon_r = 2.2$ or 10.4), Epsilon ($\varepsilon_r = 10–13$), Alumina ($\varepsilon_r = 9.6–10.4$), Fused quartz ($\varepsilon_r = 3.8$), Rutile (TiO_2) ($\varepsilon_r = 100$), and ferrite/garnet ($\varepsilon_r = 13–16$). The anisotropic substrates are sapphire ($\varepsilon_r\perp = 9.4$, $\varepsilon_r\| = 11.6$) and Pyrolytic boron nitride ($\varepsilon_r\perp = 3.4$, $\varepsilon_r\| = 5.12$).

Monolithic MICs use semiconductor substrates such as Si ($\varepsilon_r = 11.7–12$), GaAs ($\varepsilon_r = 12–13$), and sapphire ($\varepsilon_r = 11.6$, C-axis). Gallium Arsenide (GaAs) has been used extensively in the development of MMICs because of its suitability for both high-frequency transistors and low-loss passive components. MMICs predominantly use GaAs because (i) GaAs has higher saturated electron velocity and low-field mobility than silicon (ii) resulting in faster devices and (iii) GaAs can readily be made with high resistivity, making it suitable for microwave passive components.

2.11.5 Fabrication of Hybrid MICs

The fabrication of hybrid MICs involves forming the passive circuit pattern on an inert dielectric substrate and then attaching active and passive devices to the passive circuitry. The quality of the MICs, in addition to depending on substrate properties, is critically influenced by factors such as adhesive strength of the conducting pattern, homogeneity of the conductor, and the geometry definition.

Preparation of MICs starts with the selection of dielectric substrate. The choice of the substrate depends on the type of transmission line to be used. For passive circuits like circulators and isolators, the stripline can be used. If the circuit includes any active or discrete passive components like GaAs FET and chip capacitors, microstrip or slotline is more convenient. When microstrip is used, the substrates should have a higher dielectric constant to minimize radiation losses.

The metallization is carried out by hot pressing a thin foil of copper during the manufacturing process. For some substrates like ceramic or glass, which are used extensively for microstrip circuits, metallization is carried out in two steps and there are two or more metallic layers. The first thin layer is chromium and it is obtained by vacuum evaporation. This layer is about 200-AU thick and it is used to improve the adhesiveness of the subsequent layer to the ceramic. This layer is followed by a thicker layer (0.5 mil) of copper or gold. This layer is constructed by a combination of evaporation followed by electro plating.

Conductor patterns for MICs are obtained by selective photo etching of metallized substrates. First, the layout of the circuit is drawn and a photo mask is prepared. The size of the mask is equal to the actual size of the circuit. This mask is transparent at the portions where the metallization is required. Then, the substrate is cleaned and a thin layer of photoresist emulsion spread evenly over it.

Then this layer is exposed to UV light through the circuit make, due to which the exposed portions corresponding to the circuit pattern get hardened. The unexposed photoresist is dissolved away and etching of the metal is carried out. Required metallization pattern is protected by hardened photoresist and thus left unetched on the substrate. This photoresist layer on the final circuit pattern is then removed.

The next step after the photo etching of the metallized substrate in the process of obtaining the required circuit pattern is to mount the discrete (active or passive) devices. The mounting of series-connected devices may be easily achieved by bonding the devices across a gap in the microstrip. Shunt mounting on the microstrip circuits is carried out by drilling a hole through the substrate. The device is mounted between the ground plane and a strap placed over the microstrip to bridge the hole. Bonding techniques for connecting device leads to microstrips depend on the circuit metallization and type of the device package used. Generally, soldering is used with the copper-clad laminates, which are used mainly for stripline circuits, and thermo-compression bonding is used for gold metallization, which is used for microstrip and slot line circuits.

2.11.6 Advantages of MICs

Since the users of microwave spectrum are overcrowded, it is high time to use millimeter-wave frequencies. Because of the smaller wavelengths, millimeter waves offer several advantages over microwaves. MIC technology is preferred over the conventional microwave circuits for the following reasons:

i. Reduction in cost
ii. Improvement in performance
iii. Good reproducibility and reliability
iv. Small size and weight (miniaturization)
v. Wide bandwidth
vi. Flexibility in design and high packaging density

2.11.6.1 Reduction in Cost

MICs are semiconductor devices in the chip form, as chips are much cheaper than the packaged devices. The photo-etching technique, a much quicker and cheaper process, can be used. The cost of chip fabrication is reduced when produced in large quantities. However, setting up on the MMIC fabrication facility for mass production is prohibitively expensive, that is, a major initial investment is required for starting an MMIC laboratory with extensive CAD facilities. Once the circuit has been perfected and the mask prepared, large-scale production does not require any additional tooling, etc. The low-cost advantages stem from the fact that a single wafer can have more than 1,000 components, each with the same performance and automated assembly work and requiring no hand tuning.

2.11.6.2 Improvement in Performance

The circuit performance may be improved by adopting the MIC technique.

a. *Improvement in device–circuit interface*: In conventional microwave circuits, the transverse dimensions of the circuit are much larger than those for the semiconductor devices used. This necessitates the use of mounting structures, which introduces additional reactances and limits the circuit performance. In MICs, transverse dimensions of the microwave structures are compatible with the size of device chips. The package and the mounting reactances are eliminated. Elimination of these reactances increases the circuit bandwidth and also the device impedance level. Ultra-broadband P intrinsic N (PIN) diode switches have been obtained in stripline and microstrip configurations only.
b. *Use of lumped elements*: Lumped element resonators have a wider bandwidth than the corresponding distributed element circuits. This is caused by the fact that the equivalent L and C of the distributed circuits are not constant with frequency. Thus, the use of lumped

inductors allows one to have, for example, a wider bandwidth in tunnel diode amplifiers and an increased tuning range in varactor-tuned circuits.

c. *Use of special circuits combing microstrip and other lines*: The combination of microstrip and slotline in the same circuit introduces another degree of flexibility in circuit design. An increase in bandwidth of couplers has been reported by using such a combination. Similarly, slotline and coplanar lines have been combined usefully to design novel circuit configurations for balanced mixers.

2.11.6.3 Good Reproducibility and Reliability

Reproducibility and reliability are excellent for monolithic MICs because the active and passive components are produced in large quantities with a very similar performance by the same well-controlled fabrication steps using the same photolithographic masks. However, the hybrid MICs suffer from device placement and wire bonding variations from circuit to circuit. In millimeter-wave circuits, the ability to control interconnection parasitics presented by bond wires is a major advantage. Microwave semiconductor devices used in MICs have a longer life as compared to that of tubes.

2.11.6.4 Small Size and Weight

Miniaturization is a key feature of MICs. The following factors contribute to making the MICs smaller and lighter:

a. *Reduced cross section for transmission lines*: An X-band waveguide is 1 in. by 0.5 in. in size, whereas the cross-section area required for a 50 Ω microstrip on alumina substrate is of the order of 0.25 in.. Microstrip dimensions are smaller than the standard coaxial lines.
b. *Reduced guide wavelength*: The use of high dielectric constant substrates reduces the guide wavelength by a factor equal to the square-root of effective dielectric constant. For alumina circuits, a reduction of 0.2–0.3 is obtained. This reduces the size of circuits using distributed elements in the same ratio.
c. *Use of lumped elements*: A further reduction in the size is achieved by the use of lumped elements in place of distributed circuits. Dimensions of lumped elements are necessarily less than about one-tenth of the wavelength. Thus, the lumped element circuits are about an order magnitude smaller than distributed circuits.
d. *Elimination of interconnection and flexibility in layout*: Since complicated circuits' functions cannot be built in single units when waveguides or coaxial lines are used, a conventional subsystem consists of a number of separate circuits connected together by coaxial connectors, waveguide flanges, or adapters. In MICs, even a large complicated circuit

function can be produced by photo etching of a single metallization and a considerable volume occupied by connecters, etc., is saved. Also, this fabrication process allows the designer to locate the input–output ports at the desired location. Thus the amount of hardware, which conventional microwave circuits use for bends, etc., is saved.

e. *Use of device chips*: In MICs, it is possible to use encapsulated device chips. These device chips occupy much smaller space than packaged devices. Also, the mounting structure for packaged devices may require additional space.

2.11.6.5 Wide Bandwidth

Wide bandwidth amplifiers are commonly used components in most RF and microwave instrumentation and communication systems. The use of both passive and negative feedbacks help to extend the bandwidth of microwave amplifiers. Devices like GaAs MESFET were developed with special emphasis on reduction of parasitics. Experimental amplifier modules exhibit a bandwidth of more than five octaves covering a frequency band from 350 MHz to 14 GHz. The amplifier makes use of frequency-controlled feedback and simple matching techniques. Using the techniques of direct coupling schemes and with matching circuits, MMIC amplifiers with 10 dB gain over a 10 GHz bandwidth are realized. In addition to these, distributed amplifiers have been known to be capable of producing high-gain wideband performance.

2.11.7 Difficulties of MICs

The physical and technical difficulties with microwave integrated circuits are listed as follows:

i. Power-handling capacity is poor due to corona discharge in pulse operation and due to heating in continuous wave operation.
ii. The operational frequency range is limited due to the onset of parasitic waves (surface wave) at higher frequencies. The range can be extended by use of various techniques inclusive of placing absorber materials in the circuit housing.
iii. Circuits with bulky or moveable components such as vacuum tubes cannot be put to use.
iv. High isolation cannot be achieved between adjacent circuit components on the same substrate.
v. Variation of dielectric properties in many substrate materials leads to variations in the circuit parameters.
vi. MICs have the low value of Q-factor associated with microstrip resonators (of the order of 100) compared with the waveguide resonators (of the order of 1,000).

vii. Once an MIC has been fabricated, there is hardly any provision like tuning screws, variable shorts, etc. for adjustment to improve the circuit parameters. Before the circuit is fabricated, the CAD technique can be used by taking all the parameter variations into account for the accurate design of MICs.

Due to these limitations, certain special components such as transmitter output stages for RADAR, high-power attenuators, tube transmitter, etc. cannot be produced as MICs.

2.11.8 Applications of MMICs

The applications with MMICs are summarized in Table 2.5.

TABLE 2.5

Applications of MMICs

Military	Space	Civil
Phased-array radar	Communications satellites	Satellite TVRO
	Remote sensing	VSAT earth terminals
Electronic warfare	Synthetic aperture radar	Mobile phones
		LOS communications
Smart munitions	Radiometers	Wireless LANs
Remote sensing		Fiber-optic systems
Synthetic aperture radar	Astronomy	Global positioning system (GPS)
		Smart cards/tagging
Decoys	Low earth orbit Satellites (IRIDIUM)	Search and rescue transponders
		M(3)VDS and the wireless local loop
Altitude meters	Steerable phased-array antennas	Anti-collision radar
		Automatic tolling
Instrumentation		Medical systems

2.11.9 Substrates for MICs

- Dielectric constant (ε_r)
  Higher $\varepsilon_r \Longrightarrow$ smaller guide wavelength $\Longrightarrow$ smaller circuit size.
- Dielectric loss tangent ($\tan \delta_d$)
  Should be as low as possible.
  Typically, values of the order of 5×10^{-4} or less are desirable.
- Surface finish
  Higher the frequency, better must be the surface finish.
  Necessary for reducing losses and for attaining good line definition.
  Typical surface finish <1 μm at X-band.
- Substrates
  E.g.: Soft copper-clad laminates –RT-Duroid

ε_r = 2.2, 2.33, 2.94, 6; tan δ_d at 10 GHz ~ 0.0012
Standard thickness = 1/64″, 1/32″, 1/16″, 1/8″
Ceramic substrates—Alumina (99.5%)
ε_r = 9.5–10; tan δ_d at 10 GHz ~ 0.0003
Standard sizes = 1″ × 1″–4″ × 4″; thickness = 0.254, 0.635, 1.27 mm.

2.11.10 Microwave Integrated Circuits – Salient Features

Microwave circuits realized using planar techniques and technology
Elements of MIC

- Planar transmission lines
- Distributed elements
- Lumped elements
- Solid-state devices
- Other components—dielectric resonators/ferrite discs.

Fabrication

- Hybrid technology – MIC
- Monolithic technology – MMIC
- Microelectromechanical Systems – MEMS.

2.11.11 Lumped Elements for MICs

Advantages of lumped elements over distributed elements

- Smaller size
- Multilevel fabrication process
- Impedance transformers with a ratio as high as 20:1 are possible
- Useful in broadband applications
  i. RF chokes using lumped inductors
  ii. Lumped inductors for tuning active device capacitance.

2.12 Static –TEM Parameters

- Microstrip synthesis problem consists of finding the values of width (W) and length (L) corresponding to the characteristic impedance Z_0 and electric length θ.

- The synthesis actually yields the normalized width-to-height ratio $\frac{w}{h}$ initially, as well as the effective microstrip permittivity ε_{eff} (a quantity unique for mixed dielectric transmission line system and it provides a useful link between various wave lengths (λ), impedances (Z), and propagating velocities).
- The characteristic impedance Z_0 for any TEM-type transmission line at high frequencies,

$$Z_0 = \sqrt{\frac{L}{C}} = \sqrt{\frac{L \times L}{C \times L}} = \frac{L}{\sqrt{LC}} = v_p L \text{ (or) } \frac{1}{v_p C} \quad (2.36)$$

$$\beta = \omega\sqrt{LC} \Rightarrow \beta = \frac{2\pi}{\lambda_g} \quad (2.37)$$

$$v_p = \frac{\omega}{\beta} = \frac{\omega}{\omega\sqrt{LC}} = \frac{1}{\sqrt{LC}} \quad (2.38)$$

$$v_p = \frac{1}{\sqrt{\mu\varepsilon}} = \frac{c}{\sqrt{\mu_r\varepsilon_r}} = \frac{c}{\sqrt{\varepsilon_r}} \ (\because \mu_r = 1 \text{ for non ferromagnetic material}) \quad (2.39)$$

$$\lambda_g = \frac{\lambda}{\sqrt{\varepsilon_r}} \quad (2.40)$$

Removing microstrip material and replacing with air

$$Z_{01} = \sqrt{\frac{L}{C_1}} \ (\because \text{L remains unaltered}) \quad (2.41)$$

$$Z_{01} = CL = \frac{1}{CC_1} \quad (2.42)$$

Combining Z_{01} and Z_0

$$Z_0 = \frac{1}{c\sqrt{CC_1}} \quad (2.43)$$

That is, Z_0 can be obtained if C without air and with air is calculated.

2.12.1 Static Analysis

- Static analysis produces transmission line parameters that are frequency independent (Figure 2.19).

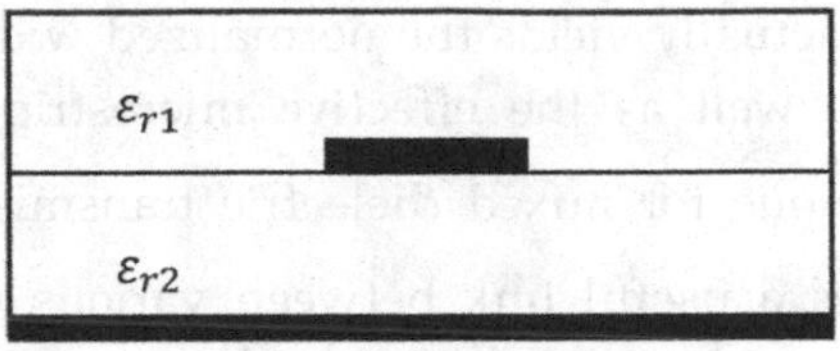

FIGURE 2.19
Static analysis of microstripline.

$$Z_0 = \sqrt{\frac{L}{C}} = \sqrt{\frac{L_a C_a}{C_a C}} = \frac{1}{c\sqrt{CC_a}} \tag{2.44}$$

L, C – Inductance, Capacitance/unit length of the transmission line.

L_a, C_a – Inductance, Capacitance/unit length of the transmission line when the dielectric is replaced by air.

Inductance/unit length does not depend on the surrounding substrate.

C = velocity of light in free space.

The phase velocity ν_p of the quasi-TEM wave propagating along the transmission line is given as

$$\nu_p = \frac{c}{\sqrt{\varepsilon_{eff}}} \tag{2.45}$$

$$\varepsilon_{eff} = \frac{c^2}{\nu_p{}^2} = \frac{LC}{L_a C_a} \tag{2.46}$$

$$\text{Wavelength } \lambda = \frac{\lambda_0}{\sqrt{\varepsilon_{eff}}} \tag{2.47}$$

Laplace equations are to be solved for "C" calculation, that is, we need to calculate capacitance of transmission line with and without the substrate.

2.13 Effects of Discontinuities

1. Frequency shift in narrow band circuits.
2. Degradation in input and output voltage standing wave ration (VSWR).

3. Higher ripple in the gain flatness of broadband ICs.
4. Interfacing problem in multifunction circuits.
5. Lower circuit yield due to degradation in circuit performance.
6. Surface wave and radiation couplings that may cause oscillations in high-gain amplifiers.

2.14 Applications of Transmission Line more than 100 GHz

2.14.1 Open Homogeneous Dielectric Guides

- Slab dielectric guides
- Rectangular dielectric guide
- Circular dielectric guide
- Because of the absence of conducting boundaries, the EM fields in these exist both inside and outside the dielectrics
- The relative amount of energy propagation inside the dielectric increases with an increase in the dielectric constant of the guide for a given cross-sectional area.

2.14.2 Image Guide

- The simplest form of the dielectric guide.
- A dielectric strip in intimate contact with a ground plane (Figure 2.20).

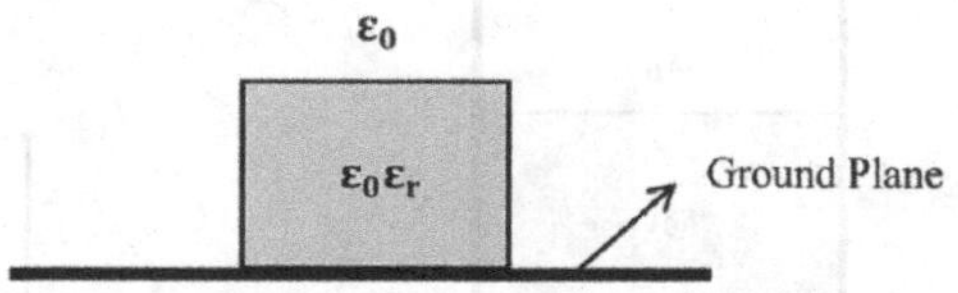

FIGURE 2.20
Image guide.

2.14.3 Nonradiative Dielectric Guide

See Figure 2.21.

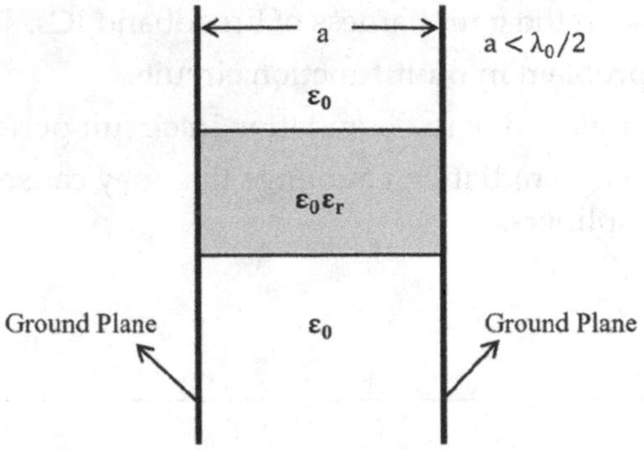

FIGURE 2.21
Nonradiative dielectric guide.

- The undesirable radiation at the bends and other discontinuities normally encountered in image guide and other open dielectric guides are suppressed.
- For use of the higher end of the microwave.
- H-guide resembles that of the non radiative guide except that the plate separation is greater than a wavelength.
- Makes use of surface wave guidance at the dielectric interface in one transverse direction and field confinement by parallel plates in the other.
- Supports H-modes with both E and H lines having a component in the direction of propagation.
- **Interesting Feature**: No longitudinal current flow on the metal walls due to the absence of a vertical component of the magnetic field at the walls (Figure 2.22).

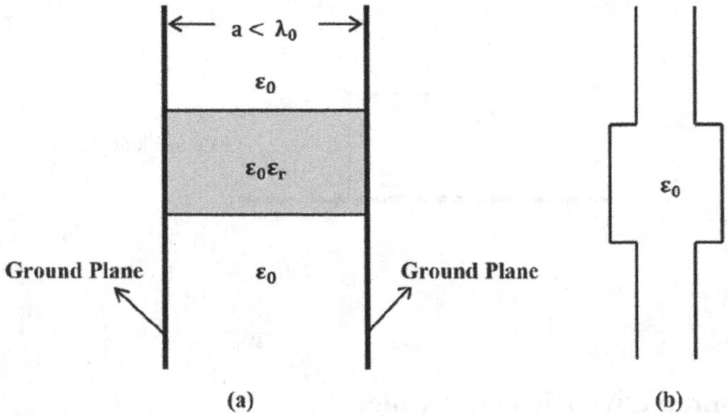

FIGURE 2.22
(a) H-guide and (b) groove guide.

2.14.4 H-Guide

- The electric field lines are essentially parallel to the conducting walls and the magnetic field lines are parallel to the dielectric surface.
- This mode offers low propagation loss, which decreases with an increase in plate separation.
- Potential over the frequency range from 100 to about 200 GHz.
- Operation beyond 200 GHz is limited because of multimode propagation. This problem is overcome in the **groove guide**.

2.14.5 Groove Guide

- Groove region creates a surface wave effect and supports slow wave propagation.
- The groove dimensions are convenient to handle at operating frequencies from 100 to 300 GHz.
- The guide can offer a single-mode operation with low propagation loss.

2.14.6 Dielectric Integrated Guide

- Low loss
- Light weight
- Ease of fabrication at frequencies in the range of 30–120 GHz
- Possibility of integration of high-performance dielectric antenna with the RF frontend.

2.15 Summary

2.15.1 Essentials of Microwave Integrated Circuit Component Design and Planar Transmission Lines

Abstract: Planar transmission lines like stripline, microstripline, slot line, coplanar wave guide (CPW), and finline form the backbone of the MICs. Today's state of the art in communication, space research, biomedical engineering, and defense advancements would not have come but for microwave engineering. But students are averse to studying the most important and indispensable subject. The material details the methods to make the students feel at ease with the subject. In short, "microwaves are made easy" by following a few procedures. RF MEMS also can be understood if and only if one is strong in the fundamentals of microwave engineering, microelectronics, microwave integrated circuits, and planar transmission lines. The RF MEMS application in biomedical engineering is studied under the name Bio MEMS. Modified microstrips and coplanar wave guides are the popular choices as transmission lines in RF MEMS. An attempt is made to have an overview of the topics S-Parameters, Microwave Circuit Theory, Planar Transmission Lines Essentials including the modes of propagation, Equivalence of Network Theory to Distributed Theory, Realization of Inductor and Capacitor in Distributed Theory, Smith Chart and its Parts in MIC Design, Smith Chart as An Art, Transformations, Microwave Integrated Circuit Component Design Overview, CAD approach, Design Essentials of the MIC Components, Matching Networks, Filters, Coupled Theory and classic References for each topic. In short, it will be an eye opener for the students and researchers to create interest in the subject of importance.

2.15.2 Introduction

From the expression of skin depth and field expressions, it is possible to show that the conducting wires are of no use at microwave frequencies, and waveguides become transmission media. From the dominant mode definition, one can see that waveguides also are of no use beyond certain frequencies and under certain conditions. Here comes the MIC where planar transmission lines (transmission lines that consist of conducting strips printed on surfaces of the transmission lines' substrates) form the backbones of it. The progress of MICs would not have been possible but for the advances of planar transmission lines. To completely characterize the behavior of transmission lines for RF and microwave designs, it is enough if few necessary properties of them are known. To know the properties of each transmission line, it is sufficient to know which mode is supported by each transmission line, which will then facilitate expanding the characteristic properties. A planar configuration implies that the characteristics of the element can be determined by the dimensions in a single plane. From the equation for the sending end impedance, $Z_s = Z_0(Z_R \cosh\gamma l + Z_0 \sinh\gamma l)/(Z_0 \cosh\gamma l + Z_R \sinh\gamma l)$, it is possible to show the miracles of "transmission line." Length in lambda makes all the differences in obtaining various microwave components and systems. A shorted small (less than quarter wave length) transmission line acts as an inductor, while that of an open line acts as a capacitor. Using the stepped impedance method, Inductor and Capacitor can also be realized. This concept felicitates all MIC filter elements, amplifiers (input and output matching networks), and oscillators (matching networks) design. While the design relations remain constant, the effective dielectric constant alone varies with various transmission lines. It is a relative dielectric constant for stripline, and for all other planar transmission lines, it is an effective dielectric constant. This makes the design easier. Low cost, small size and weight, conformability, improved reliability and reproducibility, multi-octave performance and circuit design flexibility, and multifunction performance on a chip are the major features of planar transmission lines (Figure 2.23).

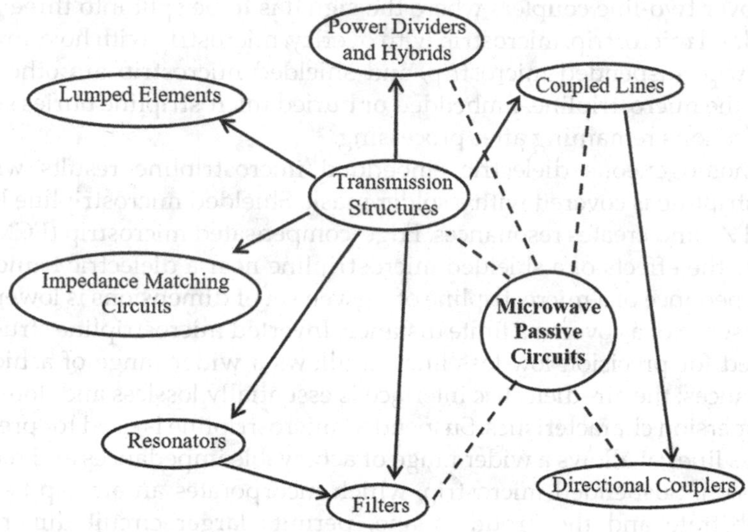

FIGURE 2.23
Concept map for passive microwave circuits.

2.15.3 Stripline and Its Variants

First in the planar transmission line family is the stripline, which is homogeneous and has TEM mode as the dominant mode of propagation. The basic structure of the homogeneous strip transmission line consists of a flat strip conductor situated symmetrically between two large ground planes, first proposed by Barret (1951) has pure TEM mode as the dominant mode of propagation. The suspended stripline, the most useful variant of the stripline (inhomogeneous transmission line), has the effective dielectric constant close to that of air. Edge-coupled suspended substrate lines have a lower loss and less sensitivity to physical dimensions than an equivalent microstrip or stripline. Shielded stripline corrects for the effect of side walls at a finite distance.

2.15.4 Microstripline and Its Variants

Microstripline, proposed by D.D. Greig and H.F. Engelmann (1952), is an inhomogeneous structure, and quasi-TEM mode is the dominant mode of propagation. Inverted microstrip, suspended microstrip, microstrip with overlay are few of the variants of microstripline. Coupled asymmetric microstriplines on ferrite substrate structures are of interest in the design of several gyromagnetic device applications. Broadside coupled microstripline is useful for tight coupling directional couplers and differential transmission of signals. Three coupled microstriplines have the potential to reduce space over two-line couplers where the signal is to be split into three paths. Embedded microstrip, microstrip with overlay, microstrip with hole, inverted microstrip, suspended microstrip, and shielded microstrip are other variants of the microstripline. Embedded or buried microstripline buries surface imperfections remaining after processing.

Nonhomogeneous dielectric embedded microstripline results when a microstripline is covered with a solder mask. Shielded microstripline lowers ε_{eff} and Z_0, and creates resonances. Edge-compensated microstrip (ECM) line corrects the effects of a shielded microstripline near a dielectric truncation. The impedance of a microstripline of a given set of dimensions is lowered by the presence of a cover at a finite distance. Inverted microstripline structures are used for precision low-loss lines; it allows a wider range of achievable impedances; the air–dielectric interface is essentially lossless and dominates the dispersion characteristics. Suspended microstripline is used for precision low-loss lines; it allows a wider range of achievable impedances and reduced dispersion. Suspended microstrip, which incorporates an air gap between the substrate and the ground plane, permits larger circuit dimensions, leading to less stringent mechanical tolerances and an increased accuracy of circuit fabrication as compared with microstrip.

2.15.5 Slotline and Its Variants

Third in the family of planar transmission lines is the slot line. Slot line, proposed by Seymour B. Cohn in 1968, consists of a narrow gap in a conductive coating on one side of a dielectric substrate, the other side of the substrate being bare. Slot line proposed by S.B. Cohn (1969) is complementary to the microstrip structurally. Non-TEM mode is the mode of propagation. While the characteristic impedance decreases with the increase in microstrip width, the characteristic impedance width increases with increase in slot width. Compared with microstrip and stripline, the Q-factor of slot line is less, around 100. Shunt mounting is easier in the slotline, whereas series connection is easier in the microstripline. The characteristic impedance varies considerably with frequency in slot line, whereas in microstrip, it exhibits only a very small variation with frequency. The effective dielectric constant of the slotline is lower than that for the microstripline for the same substrate. Coupled slot lines are strongly related to the CPW structure and the slot line structure. For very wider center conductor or loose coupling between the slots, the structure becomes two uncoupled slotlines or CPW.

2.15.6 Coplanar Waveguide and Its Variants

The coplanar waveguide, proposed by C.P. Wen in 1969, is essentially a coupled slotline, having a hybrid mode of propagation, the magnetic field in the slots at the air–dielectric interface is elliptically polarized and that property is useful in the design of nonreciprocal ferrite devices. Both shunt and series connections of active and passive devices. It eliminates the need for wraparound and via holes. The characteristic impedance is determined by the ratio of a/b, so size reduction is possible without limit. The characteristic impedance is relatively independent of the thickness of the dielectric substrate. Micro-coplanar stripline or nonsymmetrical CPW have unequal gaps. Grounded CPW with ground: Propagates in three modes, namely, microstrip, CPW, and coupled slotlines. Asymmetric CPW reduces the line impedance of symmetric CPW. Broadside coupled CPW, which is of multiple dielectric configuration, is useful in integrated circuits. Edge-coupled CPW has all of the advantages of the planarity of the coplanar structure, together with an increased possible isolation over microstripline structures.

2.15.7 Coplanar Stripline and Its Variants

Coplanar strips, the complementary CPW, are suitable for microwave monolithic integrated circuits. Like CPW, in CPS also series and shunt mounting of devices is possible. The CPS is a balanced transmission line that facilitates balanced mixers and feed network for printed dipole antennas. It offers much higher impedance ranges than that of microstriplines. Applications of CPWs include MEMS-based switches and phase shifters, high-T_c superconducting transmission lines, tunable devices using ferroelectric materials, photonic bandgap structures, and printed circuit antennas. The lower ground plane of conductor-backed CPW provides mechanical strength and acts as a heat sink. Rectangular-shaped microshielded line, V-shaped microshielded line, circular-shaped microshielded line, edge-coupled CPW without a lower ground plane, broadside-coupled CPW are the other variants of CPW. Coupled transmission lines have applications like filters, directional couplers, interdigital capacitors, and planar spiral inductors.

2.15.8 Finline and Its Variants

Paul Meir's finline was specifically useful for integrated circuits using planar techniques in an E-plane construction at millimeter-wave techniques. Finline, introduced in 1972 by Paul Meier, has low-loss performance, has less stringent tolerance requirements, is compatible with waveguides, and its ease of E-plane integration with other forms of transmission lines has been recognized as an important transmission medium for millimeter-wave integrated circuits. Finline is a suitable transmission medium for frequencies in the range of 30–100 GHz. The guide wavelength is longer than in a microstrip. It has low-loss propagation. It is useful for passive and active integrated circuits. Unilateral, bilateral, insulated, antipodal, edge-coupled unilateral, edge-coupled bilateral, double-dielectric bilateral (fins facing side walls), and double-dielectric bilateral (fins facing each other) are few of the finline variants.

2.15.9 Dielectric Guides

Dielectric guides have low transmission loss at millimeter-wave frequencies. Dielectric guides backed by metallic ground planes are more practical for integrated circuit applications. Dielectric integrated guides are useful in high mm-wave frequencies where finlines pose fabrication difficulties because of increasingly stringent dimensional tolerances. Image guide, insular image guide, trapped image guide, trapped insular image guide are the basic dielectric integrated guides; nonradiative dielectric guide, nonradiative insular dielectric guide are the insular counterparts of image guide; inverted strip dielectric guide, cladded dielectric image guide, hollow image guide, ridge dielectric guide are the variants of image guide and integrated waveguide technology.

Planar transmission lines, in particular, have been developed in response to the need for accurate analysis and design of MICs. These analysis methods paved way for further investigation and development of new planar transmission lines. Commonly used techniques for analyzing microwave and millimeter-wave planar transmission lines, in particular, and passive structures, in general, are Green's function, conformal mapping, variational, spectral domain, mode-matching methods, FEM, FDTD, method of moments (MOM), and method of lines (MOL). Multilayer planar transmission lines are especially attractive for MICs due to their flexibility and ability to realize complicated circuits, ultimately allowing very compact, high-density circuit integration. They also allow their substrates for achieving ultra-compact MICs. Furthermore, multilayer transmission lines have significantly less cross talk and distortion via appropriate selection of dielectric layers. The analyses are all based on Maxwell's equations, in general, and wave equations and boundary conditions, in particular. While techniques can change steadily, the fundamentals always remain the same.

Static or quasi approach and dynamic or full-wave approach are the two methods of planar transmission line analysis. The first approach of producing transmission line parameters for the TEM mode are valid only at directional coupler (DC) and are frequency independent. The second approach can produce transmission line parameters not only for the TEM mode but also for hybrid modes whose parameters are functions of frequency. To equalize the odd mode and even mode phase velocities in coupled lines anisotropic substrates may be used in transmission lines. Sapphire and pyrolytic boron nitride are two popular anisotropic materials.

To know the properties of a transmission line, it is enough if one knows the mode supported by each transmission line. As seen, the stripline supports the TEM mode, and the microstripline supports the quasi-TEM mode. For non-TEM mode, slot line is the best example. Finline supports the hybrid mode.

India's global achievements in space science are proven ones. In space satellite communications where size and weight are the constraints, the planar transmission lines play a vital role.

2.15.10 Conclusion

The cross section of Planar Transmission lines and the Field Lines representation are very important. Once the E lines are drawn properly (the lines should start from the signal line and then end on the ground) H lines are drawn perpendicular to E lines. The field lines will help for the inter connection of different planar transmission lines. Effective dielectric constant 'ε_{eff}' for each transmission line differs. For Homogenous line (Stripline-Example) effective dielectric constant becomes relative dielectric constant ε_r). The guide wavelength $\lambda_g = \lambda_0 / \sqrt{\varepsilon_{eff}}$ takes care of every planar transmission line.

Bibliography

Bharathi Bhat and Shiban K. Koul (1968). *Stripline-Like Transmission Lines for Microwave Integrated Circuits,* New Age International, New Delhi.

Bharathi Bhat, Shiban K. Koul (1987). *Analysis, Design and applications of Fin Lines,* Artech Housen Inc., Norwood, MA.

K. C. Gupta, Amarjit Singh (1974). *Microwave Integrated Circuits,* Wiley Eastern Private Limited, New Delhi.

Reinmut K. Hoffmann, Harlan H. Howe, Jr (1987). *Hand book of Microwave Integrated Circuits,* Artech House, Inc., Norwood, MA.

Rainee N. Simons (2001). *Coplanar Waveguide Circuitsm Components, and Systems,* Wiley-Interscience, New York.

3

Microwave Integrated Circuit (MIC) Components

3.1 Directional Coupler

See Figure 3.1.

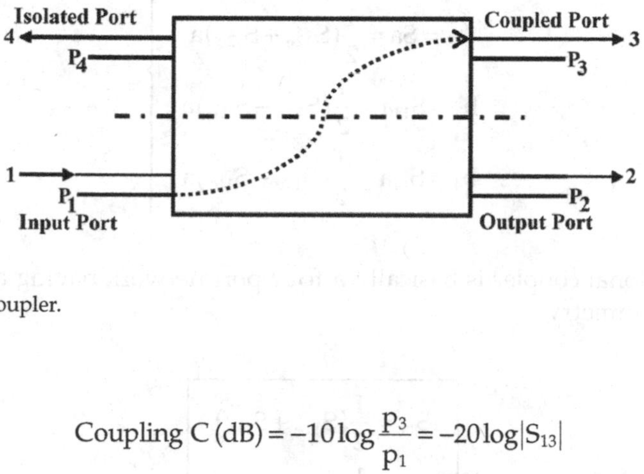

FIGURE 3.1
Directional coupler.

$$\text{Coupling C (dB)} = -10\log\frac{P_3}{P_1} = -20\log|S_{13}| \tag{3.1}$$

$$\text{Directivity D (dB)} = -10\log\frac{P_4}{P_3} \tag{3.2}$$

$$\text{Isolation L (dB)} = -10\log\frac{P_4}{P_1} = -20\log|S_{14}| \tag{3.3}$$

$$\text{Return loss R (dB)} = -10\log\frac{P_1'}{P_1} = -20\log|S_{11}| \tag{3.4}$$

See Figure 3.2.

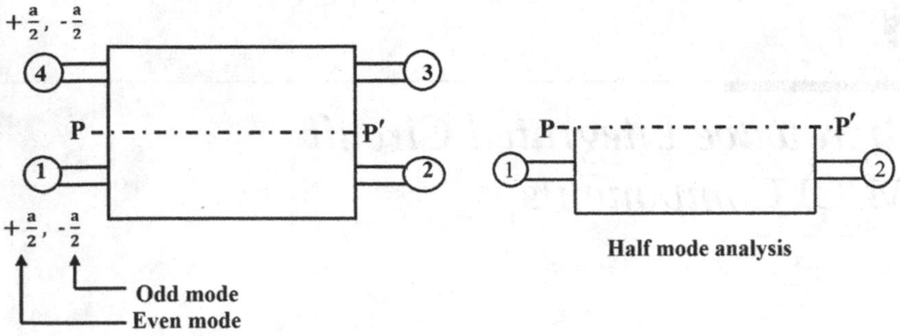

FIGURE 3.2
Half-mode analysis of directional coupler.

$$\left.\begin{aligned} b_1 &= Sa = \frac{1}{2}(S_{11e} + S_{11o})a \\ b_2 &= Sa = \frac{1}{2}(S_{12e} + S_{12o})a \\ b_3 &= S_{13}a = \frac{1}{2}(S_{12e} - S_{12o})a \\ b_4 &= S_{14}a = \frac{1}{2}(S_{11e} - S_{11o})a \end{aligned}\right\} \tag{3.5}$$

The directional coupler is basically a four-port network having at least one plane of symmetry.

$$\boxed{\begin{aligned} S_{11} &= \frac{1}{2}(S_{11e} + S_{11o}) \\ S_{12} &= \frac{1}{2}(S_{12e} + S_{12o}) \\ S_{13} &= \frac{1}{2}(S_{12e} - S_{12o}) \\ S_{14} &= \frac{1}{2}(S_{11e} - S_{11o}) \end{aligned}} \tag{3.6}$$

3.2 Two-Stub Branch-Line Coupler

See Figure 3.3.

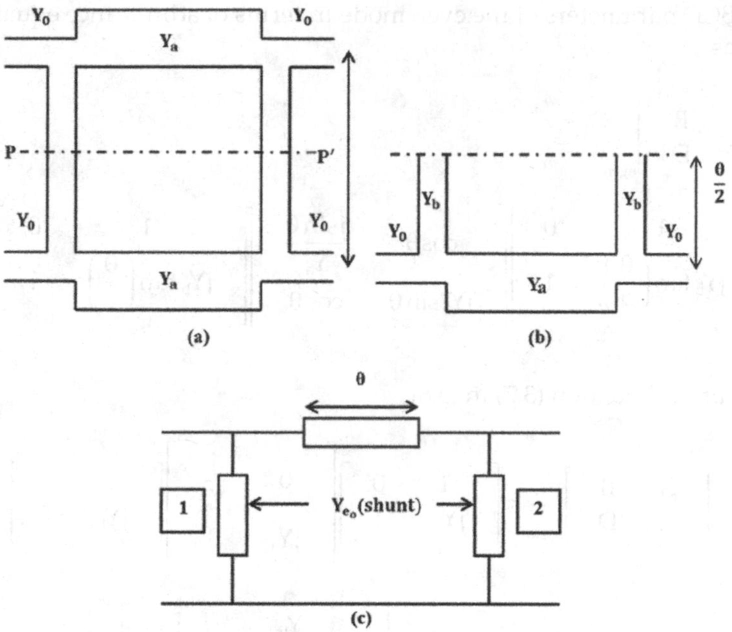

FIGURE 3.3
(a) Two-stub branch-line coupler. (b) One-half section for even- and odd-mode analyses. (c) Equivalent circuit.

where $\theta = \beta l$ and $\beta = \dfrac{2\pi}{\lambda_g}$ & $l = \dfrac{\lambda_g}{4}$. Since $\theta = \dfrac{2\pi}{\lambda_g} \times \dfrac{\lambda_g}{4} = \dfrac{\pi}{2}$

$$\left.\begin{aligned} &\text{For full-mode analysis} \quad \boxed{\therefore \theta = \frac{\pi}{2} \text{ or } 90^\circ} \\ &\therefore \text{For half-mode analysis} \quad \boxed{\therefore \frac{\theta}{2} = \frac{\pi}{4} \text{ or } 45^\circ} \end{aligned}\right\} \tag{3.7}$$

The characteristic admittance, Y_a, Y_b of the series and the shunt branch, respectively. The impedance and admittance of even mode and odd mode are given as

$$\left.\begin{aligned} &Z_{sc} = jZ_0 \tan\left(\frac{\theta}{2}\right) \Rightarrow Y_{sc} = -jY_0 \cot\left(\frac{\theta}{2}\right) (\because \text{odd mode}) \\ &Z_{oc} = \frac{Z_0}{j\tan\left(\frac{\theta}{2}\right)} \Rightarrow Y_{oc} = jY_0 \tan\left(\frac{\theta}{2}\right) (\because \text{Even mode}) \end{aligned}\right\} \text{for half mode analyzes}$$

The ABCD parameters of the even mode in terms of admittance equation are given as

$$\begin{bmatrix} A & B \\ C & D \end{bmatrix}_{even}$$

$$= \begin{bmatrix} 1 & 0 \\ jY_b \tan\left(\frac{\theta}{2}\right) & 1 \end{bmatrix} \begin{bmatrix} \cos\theta & \frac{j\sin\theta}{Y_a} \\ jY_a \sin\theta & \cos\theta \end{bmatrix} \begin{bmatrix} 1 & 0 \\ jY_b \tan\left(\frac{\theta}{2}\right) & 1 \end{bmatrix} \tag{3.8}$$

Substituting Equation (3.7) in (3.8)

$$\begin{bmatrix} A & B \\ C & D \end{bmatrix}_{even} = \begin{bmatrix} 1 & 0 \\ jY_b & 1 \end{bmatrix} \begin{bmatrix} 0 & \frac{j}{Y_a} \\ jY_a & 0 \end{bmatrix} \begin{bmatrix} 1 & 0 \\ jY_b & 1 \end{bmatrix}$$

$$\begin{bmatrix} A & B \\ C & D \end{bmatrix}_{even} = \begin{bmatrix} -\frac{Y_b}{Y_a} & \frac{j}{Y_a} \\ jY_a - j\frac{Y_b^2}{Y_a} & -\frac{Y_b}{Y_a} \end{bmatrix}$$

$$\boxed{\begin{bmatrix} A & B \\ C & D \end{bmatrix}_{even} = \begin{bmatrix} -\frac{Y_b}{Y_a} & \frac{j}{Y_a} \\ j\left(\frac{Y_a^2 - Y_b^2}{Y_a}\right) & -\frac{Y_b}{Y_a} \end{bmatrix}} \tag{3.9}$$

Similarly, for odd mode

$$Y_{sc} = -jY_0 \cot\left(\frac{\theta}{2}\right)$$

$$\begin{bmatrix} A & B \\ C & D \end{bmatrix}_{odd}$$

$$= \begin{bmatrix} 1 & 0 \\ -jY_b \cot\left(\frac{\theta}{2}\right) & 1 \end{bmatrix} \begin{bmatrix} \cos\theta & \frac{j\sin\theta}{Y_a} \\ jY_a \sin\theta & \cos\theta \end{bmatrix} \begin{bmatrix} 1 & 0 \\ -jY_b \cot\left(\frac{\theta}{2}\right) & 1 \end{bmatrix} \tag{3.10}$$

Substituting Equation (3.7) in (3.10)

$$\begin{bmatrix} A & B \\ C & D \end{bmatrix}_{odd} = \begin{bmatrix} 1 & 0 \\ -jY_b & 1 \end{bmatrix} \begin{bmatrix} 0 & \frac{j}{Y_a} \\ jY_a & 0 \end{bmatrix} \begin{bmatrix} 1 & 0 \\ -jY_b & 1 \end{bmatrix}$$

$$\begin{bmatrix} A & B \\ C & D \end{bmatrix}_{odd} = \begin{bmatrix} \frac{Y_b}{Y_a} & \frac{j}{Y_a} \\ jY_a - j\frac{Y_b^2}{Y_a} & \frac{Y_b}{Y_a} \end{bmatrix}$$

$$\boxed{\begin{bmatrix} A & B \\ C & D \end{bmatrix}_{odd} = \begin{bmatrix} \frac{Y_b}{Y_a} & \frac{j}{Y_a} \\ j\left(\frac{Y_a^2 - Y_b^2}{Y_a}\right) & -\frac{Y_b}{Y_a} \end{bmatrix}} \tag{3.11}$$

From Equations (3.9) and (3.11), the ABCD parameters of even mode and odd mode are given as

$$\boxed{\begin{bmatrix} A & B \\ C & D \end{bmatrix}_{\substack{even \\ odd}} = \begin{bmatrix} \mp\frac{Y_b}{Y_a} & \frac{j}{Y_a} \\ j\left(\frac{Y_a^2 - Y_b^2}{Y_a}\right) & \mp\frac{Y_b}{Y_a} \end{bmatrix}} \tag{3.12}$$

3.2.1 Conversion of ABCD Parameters into S-Parameter of Branch-Line Coupler

$$\left.\begin{aligned} S_{11} &= \frac{A + \frac{B}{Z_0} - CZ_0 - D}{A + \frac{B}{Z_0} + CZ_0 + D} \qquad S_{12} = \frac{2(AD - BC)}{A + \frac{B}{Z_0} + CZ_0 + D} \\ S_{21} &= \frac{2}{A + \frac{B}{Z_0} + CZ_0 + D} \qquad S_{22} = \frac{-A + \frac{B}{Z_0} - CZ_0 + D}{A + \frac{B}{Z_0} + CZ_0 + D} \end{aligned}\right\} \tag{3.13}$$

For a reciprocal network: $AD - BC = 1$

For a symmetrical network: $A = D$

The abovementioned symmetric half two-port network is symmetrical and reciprocal. From Equation (3.13),

$$\therefore S_{11e_o} = \frac{\frac{B}{Z_0} - CZ_0}{A + \frac{B}{Z_0} + CZ_0 + D} = \left[\frac{BY_0 - \frac{C}{Y_0}}{A + BY_0 + \frac{C}{Y_0} + D}\right]_{e_o} \tag{3.14}$$

$$S_{12e_o} = \left[\frac{2}{A + BY_0 + \frac{C}{Y_0} + D}\right]_{e_o} \tag{3.15}$$

In any ideal directional coupler perfectly matched at the design frequency,

$$\boxed{p' = p_4 = 0} \tag{3.16}$$

Hence, directivity, isolation, and return loss expressed in decibels must become infinitive. Further, when $S_{11} = 0$, then Equation (3.14) becomes

$$\boxed{BY_0 = \frac{C}{Y_0}} \tag{3.17}$$

Get ABCD values from Equation (3.12) and substitute in Equation (3.15)

$$S_{12e_o} = \left[\frac{2}{A + BY_0 + \frac{C}{Y_0} + D}\right]_{e_o}$$

Consider denominator

$$A + BY_0 + \frac{C}{Y_0} + D = \mp\frac{Y_b}{Y_a} + \frac{j}{Y_a}Y_0 + j\frac{\left(\frac{Y_a{}^2 - Y_b{}^2}{Y_a}\right)}{Y_0} + \left(\mp\frac{Y_b}{Y_a}\right)$$

$$= \mp\frac{2Y_b}{Y_a} + \frac{j}{Y_a}Y_0 + j\left(\frac{Y_a{}^2 - Y_b{}^2}{Y_aY_0}\right)$$

$$= \frac{\mp 2Y_b + jY_0}{Y_a} + j\frac{Y_a{}^2 - Y_b{}^2}{Y_aY_0}$$

$$= \frac{\mp 2Y_aY_bY_0 + jY_0{}^2Y_a + jY_a\left(Y_a{}^2 - Y_b{}^2\right)}{Y_a{}^2Y_0} \quad [\because \text{by cross multiplication}]$$

$$= \frac{Y_0\left[\mp 2Y_aY_b + jY_0Y_a + j\,\dfrac{Y_a}{Y_0}\left(Y_a{}^2 - Y_b{}^2\right)\right]}{Y_a{}^2Y_0}$$

Therefore, Equation (3.15) becomes

$$S_{12e_o} = \left[\frac{2Y_a{}^2}{\mp 2Y_aY_b + jY_0Y_a + j\,\dfrac{Y_a}{Y_0}\left(Y_a{}^2 - Y_b{}^2\right)}\right]_{e_o}$$

$$S_{12_e} = \left[\frac{2Y_a{}^2}{-2Y_aY_b + jY_0Y_a + j\dfrac{Y_a}{Y_0}\left(Y_a{}^2 - Y_b{}^2\right)}\right] \text{[Even mode]} \tag{3.18}$$

$$S_{12_o} = \left[\frac{2Y_a{}^2}{2Y_aY_b + jY_0Y_a + j\dfrac{Y_a}{Y_0}\left(Y_a{}^2 - Y_b{}^2\right)}\right]_o \text{[Odd mode]} \tag{3.19}$$

Also,

$$S_{11e_o} = \left[\frac{A + BY_0 - \dfrac{C}{Y_0} - D}{A + BY_0 + \dfrac{C}{Y_0} + D}\right]_{e_o}$$

If the input port is matched, then $S_{11e_o} = 0$. Applying symmetrical network condition A = D, the preceding equation becomes

$$BY_0 = \frac{C}{Y_0}$$

Substituting B and C values from Equation (3.12),

$$\frac{j}{Y_a}Y_0 = j\frac{\left(\dfrac{Y_a{}^2 - Y_b{}^2}{Y_a}\right)}{Y_0}$$

$$\Rightarrow \frac{Y_0}{Y_a} = \frac{Y_a^2 - Y_b^2}{Y_a Y_0}$$

$$\Rightarrow Y_0^2 = Y_a^2 - Y_b^2 \tag{3.20}$$

See Figure 3.4.

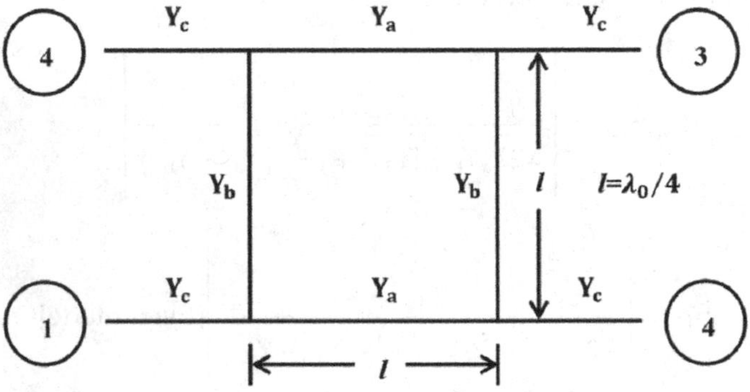

FIGURE 3.4
Design parameters of the two-stub branch-line coupler.

3.2.2 Midband Parameters

f_0: midband frequency

λ_0: midband wavelength in the medium (Table 3.1; Figure 3.5)

$$S_{11} = S_{14} = 0, \quad S_{12} = -jY_c/Y_a, \quad S_{13} = -Y_b/Y_a$$

$$Y_c^2 = Y_a^2 - Y_b^2$$

TABLE 3.1
Branch Admittances

	S_{12}	S_{13}	Y_a	Y_b
3 dB, 90° coupler	$-\frac{j}{\sqrt{2}}$	$-\frac{1}{\sqrt{2}}$	$\sqrt{2}Y_c$	Y_c
6 dB, 90° coupler	$-j\frac{\sqrt{3}}{2}$	$-\frac{1}{2}$	$\frac{2Y_c}{\sqrt{3}}$	$\frac{Y_c}{\sqrt{3}}$
10 dB, 90° coupler	$-j\frac{3}{\sqrt{10}}$	$-\frac{1}{\sqrt{10}}$	$\frac{\sqrt{10}Y_c}{3}$	$\frac{Y_c}{3}$

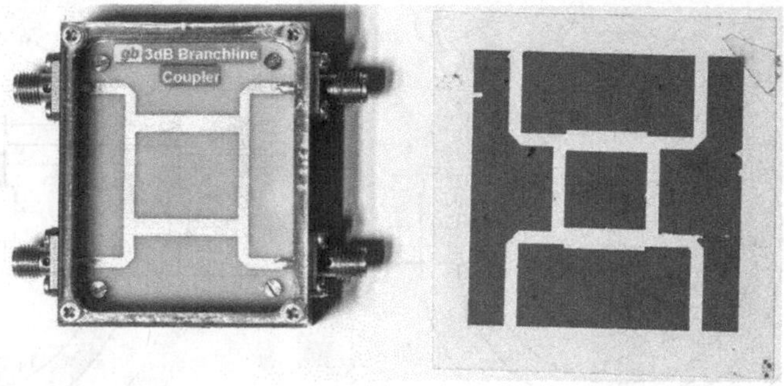

FIGURE 3.5
Fabricated structure and layout of the two-stub branch-line coupler.

3.3 Hybrid Ring Coupler

See Figure 3.6.

For symmetrical analysis, θ value = θ + 2θ + 3θ =6θ
For full analysis, θ value = 6θ + 60 = 12θ

$$\boxed{\theta = \frac{\pi}{4}}$$

$$\text{Since } 12\theta = \beta l = \frac{2\pi}{\lambda_g} \times \frac{3\lambda_g}{2} \tag{3.21}$$

The ABCD parameters for even mode symmetrical rat-race coupler are given as

$$\begin{bmatrix} A & B \\ C & D \end{bmatrix}_{even}$$

$$= \begin{bmatrix} 1 & 0 \\ jY_a \tan 3\theta & 1 \end{bmatrix} \begin{bmatrix} \cos 2\theta & \frac{j\sin 2\theta}{Y_b} \\ jY_b \sin 2\theta & \cos 2\theta \end{bmatrix} \begin{bmatrix} 1 & 0 \\ jY_a \tan \theta & 1 \end{bmatrix} \tag{3.22}$$

Substituting Equation (3.21) in (3.22), we get

$$\begin{bmatrix} A & B \\ C & D \end{bmatrix}_{even} = \begin{bmatrix} 1 & 0 \\ -jY_a & 1 \end{bmatrix} \begin{bmatrix} 0 & \frac{j}{Y_b} \\ jY_b & 0 \end{bmatrix} \begin{bmatrix} 1 & 0 \\ jY_a & 1 \end{bmatrix}$$

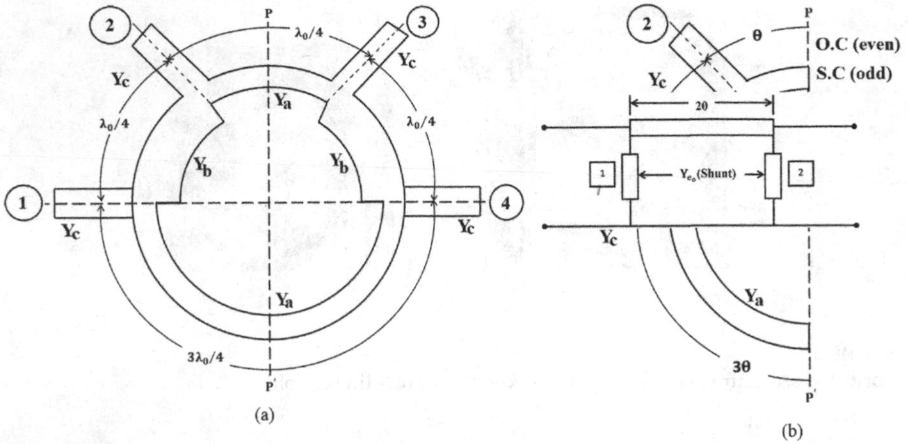

FIGURE 3.6
(a) Hybrid ring coupler and (b) one-half of the structure for the purpose of analysis and its equivalent circuit.

$$\begin{bmatrix} A & B \\ C & D \end{bmatrix}_{even} = \begin{bmatrix} -\dfrac{Y_a}{Y_b} & \dfrac{j}{Y_b} \\ jY_b + j\dfrac{Y_a^{\ 2}}{Y_b} & -\dfrac{Y_a}{Y_b} \end{bmatrix}$$

$$\boxed{\begin{bmatrix} A & B \\ C & D \end{bmatrix}_{even} = \begin{bmatrix} -\dfrac{Y_a}{Y_b} & \dfrac{j}{Y_b} \\ j\left(\dfrac{Y_a^{\ 2}+Y_b^{\ 2}}{Y_a}\right) & -\dfrac{Y_a}{Y_b} \end{bmatrix}} \tag{3.23}$$

Similarly, for odd mode

$$\begin{bmatrix} A & B \\ C & D \end{bmatrix}_{odd}$$

$$= \begin{bmatrix} 1 & 0 \\ -jY_a \cot 3\theta & 1 \end{bmatrix} \begin{bmatrix} \cos 2\theta & \dfrac{j\sin 2\theta}{Y_b} \\ jY_b \sin 2\theta & \cos 2\theta \end{bmatrix} \begin{bmatrix} 1 & 0 \\ -jY_a \cot\theta & 1 \end{bmatrix}$$

$$\begin{bmatrix} A & B \\ C & D \end{bmatrix}_{odd} = \begin{bmatrix} 1 & 0 \\ jY_a & 1 \end{bmatrix} \begin{bmatrix} 0 & \dfrac{j}{Y_b} \\ jY_b & 0 \end{bmatrix} \begin{bmatrix} 1 & 0 \\ -jY_a & 1 \end{bmatrix}$$

$$\begin{bmatrix} A & B \\ C & D \end{bmatrix}_{odd} = \begin{bmatrix} \dfrac{Y_a}{Y_b} & \dfrac{j}{Y_b} \\ jY_b + j\dfrac{Y_a^2}{Y_b} & \dfrac{Y_a}{Y_b} \end{bmatrix}$$

$$\begin{bmatrix} A & B \\ C & D \end{bmatrix}_{odd} = \begin{bmatrix} \dfrac{Y_a}{Y_b} & \dfrac{j}{Y_b} \\ j\left(\dfrac{Y_a^2 + Y_b^2}{Y_a}\right) & \dfrac{Y_a}{Y_b} \end{bmatrix} \tag{3.24}$$

From Equations (3.23) and (3.24), the ABCD parameters of even mode and odd mode for rat-race coupler are given as

$$\boxed{\begin{bmatrix} A & B \\ C & D \end{bmatrix}_{\substack{even \\ odd}} = \begin{bmatrix} \mp\dfrac{Y_a}{Y_b} & \dfrac{j}{Y_b} \\ j\left(\dfrac{Y_a^2 + Y_b^2}{Y_a}\right) & \mp\dfrac{Y_a}{Y_b} \end{bmatrix}} \tag{3.25}$$

Using the conversion of ABCD parameters into S-parameter formulas, we get

$$[S]_{even} = \begin{bmatrix} \dfrac{1 - \overline{Y_a}^2 - \overline{Y_b}^2 + j2\overline{Y_a}}{1 + \overline{Y_a}^2 + \overline{Y_b}^2} & \dfrac{-j2\overline{Y_b}}{1 + \overline{Y_a}^2 + \overline{Y_b}^2} \\ \dfrac{-j2\overline{Y_b}}{1 + \overline{Y_a}^2 + \overline{Y_b}^2} & \dfrac{1 - \overline{Y_a}^2 - \overline{Y_b}^2 - j2\overline{Y_a}}{1 + \overline{Y_a}^2 + \overline{Y_b}^2} \end{bmatrix} \tag{3.26}$$

$$[S]_{odd} = \begin{bmatrix} \dfrac{1 - \overline{Y_a}^2 - \overline{Y_b}^2 - j2\overline{Y_a}}{1 + \overline{Y_a}^2 + \overline{Y_b}^2} & \dfrac{-j2\overline{Y_b}}{1 + \overline{Y_a}^2 + \overline{Y_b}^2} \\ \dfrac{-j2\overline{Y_b}}{1 + \overline{Y_a}^2 + \overline{Y_b}^2} & \dfrac{1 - \overline{Y_a}^2 - \overline{Y_b}^2 + j2\overline{Y_a}}{1 + \overline{Y_a}^2 + \overline{Y_b}^2} \end{bmatrix} \tag{3.27}$$

where $\overline{Y} = \dfrac{Y}{Y_0}$ (normalized admittance).

$$\text{If the input port is matched then } S_{11} = S_{13} = 0 \tag{3.28}$$

But we know that

$$\left.\begin{aligned} S_{11} &= \frac{1}{2}(S_{11e} + S_{11o}) \\ S_{12} &= \frac{1}{2}(S_{12e} + S_{12o}) \\ S_{13} &= \frac{1}{2}(S_{12e} - S_{12o}) \\ S_{14} &= \frac{1}{2}(S_{11e} - S_{11o}) \end{aligned}\right\} \tag{3.29}$$

Getting S_{11e} and S_{11o} values from Equations (3.26) and (3.27) and substituting Equation (3.28) in Equation (3.29), we get

$$\frac{1}{2}\left[\frac{1-\overline{Y_a}^2-\overline{Y_b}^2+j2\overline{Y_a}}{1+\overline{Y_a}^2+\overline{Y_b}^2}+\frac{1-\overline{Y_a}^2-\overline{Y_b}^2-j2\overline{Y_a}}{1+\overline{Y_a}^2+\overline{Y_b}^2}\right]=0$$

$$\frac{1}{2}\left[\frac{1-\overline{Y_a}^2-\overline{Y_b}^2+j2\overline{Y_a}+1-\overline{Y_a}^2-\overline{Y_b}^2-j2\overline{Y_a}}{1+\overline{Y_a}^2+\overline{Y_b}^2}\right]=0$$

$$1-\overline{Y_a}^2-\overline{Y_b}^2=0$$

$$\boxed{\overline{Y_a}^2+\overline{Y_b}^2=1} \tag{3.30}$$

Getting S_{12e} and S_{12o} values from Equations (3.26) and (3.27) and substituting in (3.29),

$$S_{12} = \frac{1}{2}\left[\frac{-j2\overline{Y_b}}{1+\overline{Y_a}^2+\overline{Y_b}^2}+\frac{-j2\overline{Y_b}}{1+\overline{Y_a}^2+\overline{Y_b}^2}\right]$$

$$S_{12} = \frac{1}{2}\left[\frac{-j2\overline{Y_b}}{2}+\frac{-j2\overline{Y_b}}{2}\right] \quad [\because \text{ from Equation 3.28}]$$

$$\boxed{S_{12} = -j\overline{Y_b}} \tag{3.31}$$

Similarly, substituting even-mode and odd-mode values of S_{13} in Equation (3.29), we get

$$S_{14} = \frac{1}{2}\left[\frac{1-\overline{Y_a}^2-\overline{Y_b}^2+j2\overline{Y_a}}{1+\overline{Y_a}^2+\overline{Y_b}^2} - \frac{1-\overline{Y_a}^2-\overline{Y_b}^2+j2\overline{Y_a}}{1+\overline{Y_a}^2+\overline{Y_b}^2}\right]$$

$$S_{14} = \frac{1}{2}\left[\frac{1-\overline{Y_a}^2-\overline{Y_b}^2+j2\overline{Y_a}-1+\overline{Y_a}^2+\overline{Y_b}^2+j2\overline{Y_a}}{1+\overline{Y_a}^2+\overline{Y_b}^2}\right]$$

$$S_{14} = \frac{1}{2}\left[\frac{j2\overline{Y_a}+j2\overline{Y_a}}{1+\overline{Y_a}^2+\overline{Y_b}^2}\right] = \frac{1}{2}\left[\frac{j4\overline{Y_a}}{2}\right] [\because \text{from Equation 3.28}]$$

$$\boxed{S_{14} = j\overline{Y_a}} \tag{3.32}$$

Substituting Equations (3.28), (3.31), and (3.32) in 4 × 4 matrix for full analysis of rat-race coupler.

$$\boxed{[S]_{\text{hybrid}} = \begin{bmatrix} 0 & -j\overline{Y_b} & 0 & -j\overline{Y_b} \\ -j\overline{Y_b} & 0 & -j\overline{Y_b} & 0 \\ 0 & -j\overline{Y_b} & 0 & -j\overline{Y_b} \\ -j\overline{Y_b} & 0 & -j\overline{Y_b} & 0 \end{bmatrix}} \tag{3.33}$$

In case $Y_a = Y_b$ then, both the port widths are equal and the hybrid becomes a rat-race coupler.

$$\boxed{Y_a = Y_b = \frac{Y_0}{\sqrt{2}}} \tag{3.34}$$

It is a planar version of the magic tee. The scattering matrix of rat-race is given as

$$\boxed{[S]_{\text{Rat race}} = \frac{1}{\sqrt{2}}\begin{bmatrix} 0 & -j & 0 & j \\ -j & 0 & -j & 0 \\ 0 & -j & 0 & -j \\ j & 0 & -j & 0 \end{bmatrix}} \tag{3.35}$$

3.4 Back Waveguide Coupler

See Figure 3.7.

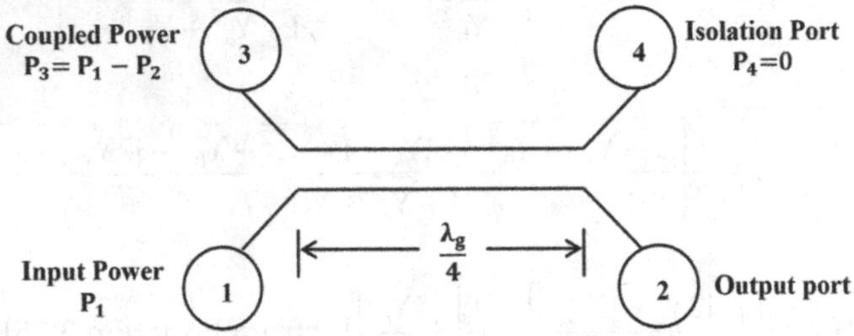

FIGURE 3.7
Back waveguide coupler.

$$Z_{even} = Z_0\sqrt{\frac{1+C_0}{1-C_0}} \quad (3.36)$$

$$Z_{odd} = Z_0\sqrt{\frac{1-C_0}{1+C_0}} \quad (3.37)$$

3.5 Basic T-Junction Power Divider

3.5.1 Scattering Matrix of Basic Power Divider

See Figure 3.8.

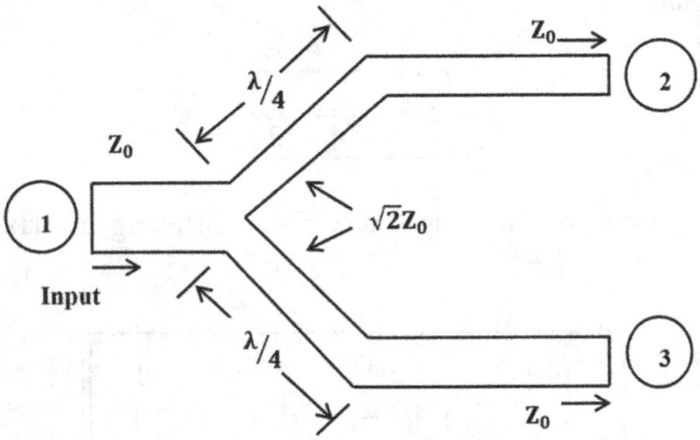

FIGURE 3.8
Basic T-junction power divider.

Assuming port 1 is perfectly matched,

$$S_{11} = 0 \tag{3.38}$$

Power fed to port 1 is equally divided between port 2 and port 3.

$$|S_{21}| = |S_{31}| = \frac{1}{\sqrt{2}} \left(\text{With this condition } S_{21} \text{ will never be equal to } 0\right)$$

Moving reference planes in branch port 2 and port 3 such that $\Phi_{21} = \Phi_{31} = 0$

(3.39)

$$\Rightarrow S_{21} = \frac{1}{\sqrt{2}}, S_{31} = \frac{1}{\sqrt{2}} \tag{3.40}$$

$$[S] = \begin{bmatrix} 0 & \frac{1}{\sqrt{2}} & \frac{1}{\sqrt{2}} \\ \frac{1}{\sqrt{2}} & S_{22} & S_{23} \\ \frac{1}{\sqrt{2}} & S_{32} & S_{33} \end{bmatrix} \tag{3.41}$$

Using the unitary condition,

$$\text{Taking second row } \frac{1}{2} + S_{22}S_{22}^{*} + S_{23}S_{23}^{*} = 1 \tag{3.42}$$

$$\text{Taking third row } \frac{1}{2} + S_{23}S_{23}^{*} + S_{33}S_{33}^{*} = 1 \tag{3.43}$$

Comparing Equations (3.42) and (3.43)

$$|S_{22}|^{2} = |S_{33}|^{2} \tag{3.44}$$

From row two, $\frac{1}{\sqrt{2}}S_{22}^{*} + \frac{1}{\sqrt{2}}S_{23}^{*} = 0$ (Figure 3.9)

$$\boxed{\therefore S_{22} = -S_{23}} \tag{3.45}$$

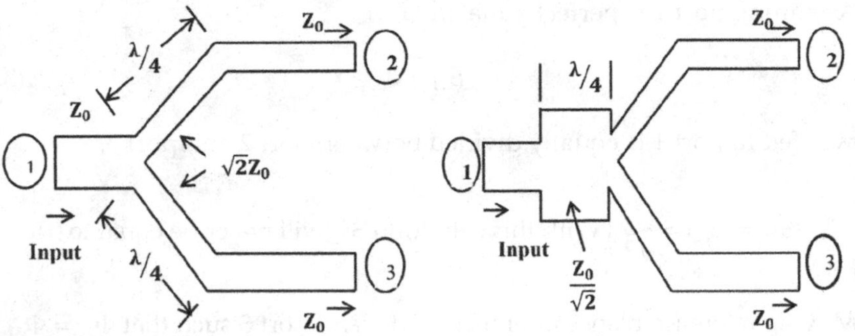

FIGURE 3.9
T-junction power divider.

Port 1 is matched: $S_{11} = 0$

$$S_{21} = S_{31}; |S_{21}| = |S_{31}| = \frac{1}{\sqrt{2}} \tag{3.46}$$

$$S_{22} \neq 0; S_{33} \neq 0$$

$$S_{23} = S_{32} \neq 0;$$

$$[S] = \begin{bmatrix} 0 & \frac{1}{\sqrt{2}} & \frac{1}{\sqrt{2}} \\ \frac{1}{\sqrt{2}} & S_{22} & S_{23} \\ \frac{1}{\sqrt{2}} & S_{32} & S_{33} \end{bmatrix} \tag{3.47}$$

Applying unitary condition $S_{22} = -S_{23}$

$|S_{22}| = |S_{33}| = |S_{23}| = \frac{1}{2}$ (ports 2 and 3 are not mutually isolated; reflection coefficient at ports 2 and 3 $= \frac{1}{2}$).

3.5.2 Matched Two-Way Power Divider

See Figure 3.10.

- Introduce a resistor $R = 2\sqrt{Z_0}$ for isolating ports 2 and 3; it is called Wilkinson power divider/combiner.
- All three ports are matched.
- Ports 2 and 3 are mutually isolated.

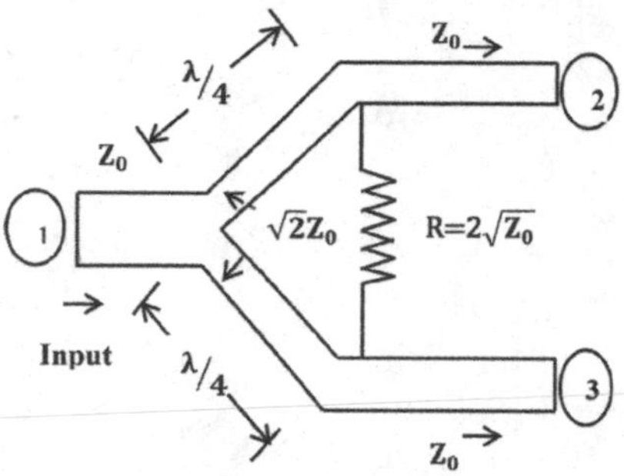

FIGURE 3.10
Matched two-way power divider.

- The device acts as a 3 dB power divider/combiner.

$$[S] = \frac{-j}{\sqrt{2}}\begin{bmatrix} 0 & 1 & 1 \\ 1 & 0 & 0 \\ 1 & 0 & 0 \end{bmatrix} \tag{3.48}$$

- Bandwidth ~ 1.44:1 for input VSWR ~ 1.22 and isolation ~ 20 dB (Figure 3.11).

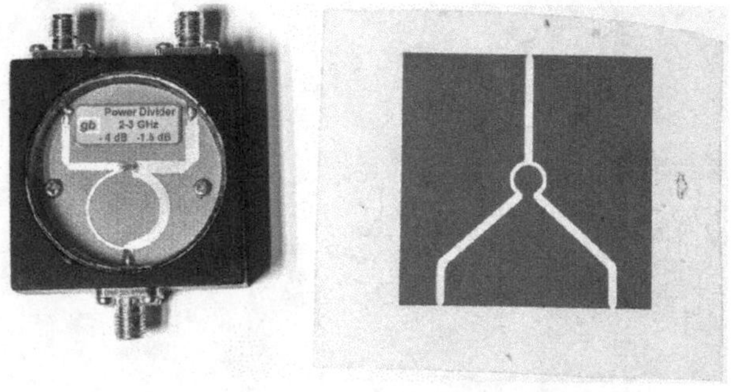

FIGURE 3.11
Fabricated structure and layout of the matched two-way power divider.

Bibliography

Bharathi Bhat and Shiban K. Koul (1968). *Stripline-Like Transmission Lines for Microwave Integrated Circuits*, New Age International, New Delhi.
Fred Gardiol (1994). *Microstrip Circuits*, Wiley-Interscience, New York.
Thomas H. Lee (2004). *Planar Microwave Engineering*, Cambridge University Press, Cambridge.
David M. Pozar (1990). *Microwave Engineering*, John Wiley & Sons Inc., New York.

4

Microwave Integrated Circuit Filters

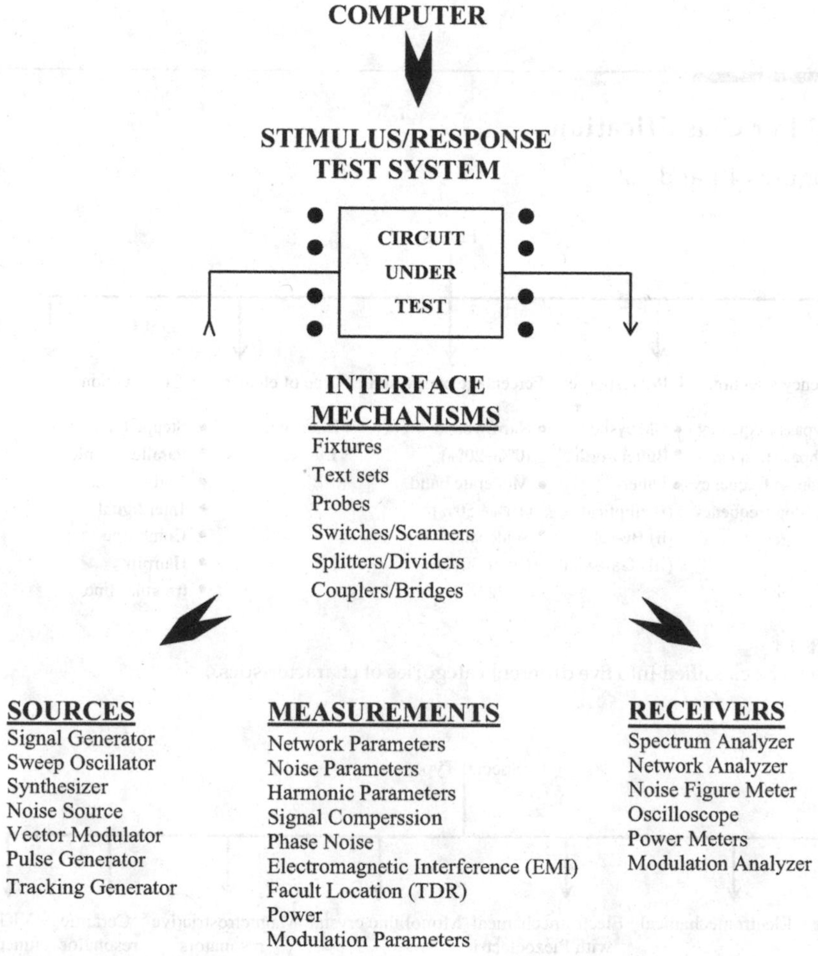

Composite automated MIC system.

4.1 Introduction

Electric filters form an essential part of all communication systems. Filtering is a fundamental process in signal processing that includes diverse applications such as channeling, demodulation, equalization, detection, decoding, phase splitting, integration/differentiation, and multiplexing. Filtering is also used in mixers for separating the radio frequency signal from the intermediate frequency. They also find applications in microwave-measuring instruments such as network analyzers.

4.2 Filter Classification

See Figures 4.1 and 4.2.

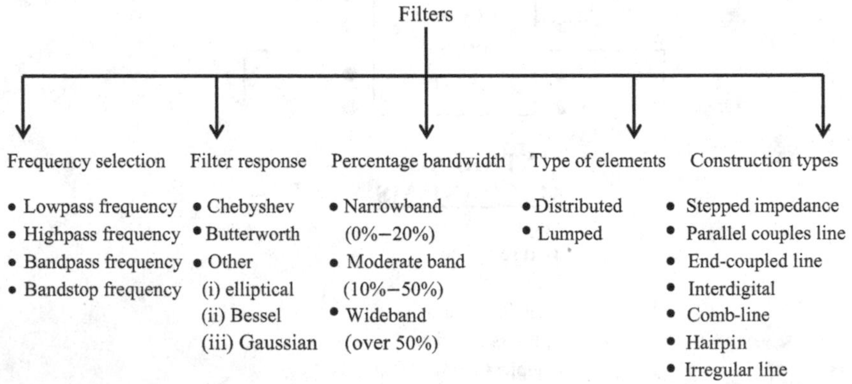

FIGURE 4.1
Filters can be classified into five different categories of characteristics.

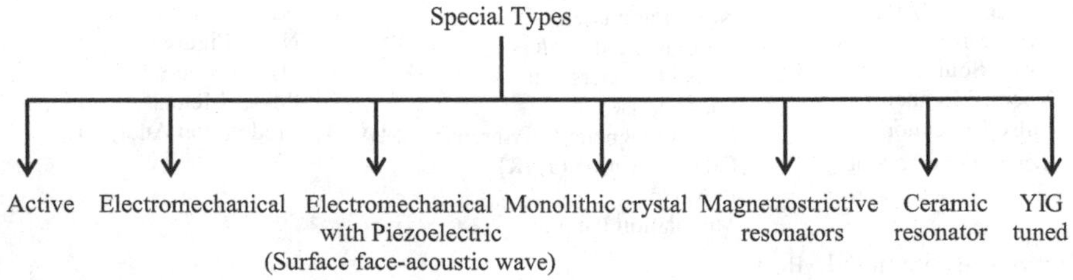

FIGURE 4.2
Special types of filters.

4.3 Coupling Matrix

Important developments in the field of microwave filter date back to the 1970s when the microwave filters were synthesized by extracting the lumped elements like inductors and capacitors, and determining the transmission line lengths. This was perfectly adequate for the state-of-the-art technologies and important applications available during the time. It was the advent of revolution in the telecommunication systems in the early 1970s that posed remarkable challenges to the microwave filter industry. This required that radio frequency spectrum allocated to the satellite communication systems be pushed to higher frequencies. This high traffic in the spectrum demanded for stringent designs and called for filters with high levels of attenuation. This demanded two important advancements in the field of filter design—filters with in-built transmission zeros, and introduction of cross-coupled filters, which allowed inter-resonator coupling. This was made possible with generalized Chebyshev filters.

The standard procedure of filter synthesis starts with first generating the polynomials representing the transfer and reflection characteristic function satisfying the filter specifications. The prototype element values are then extracted from the generated polynomials. Coupling matrix synthesis method is an alternative to the extraction of the prototype elements one by one. The coupling matrix is a representation of the network which can be used to analyze the performance of the microwave filter. The concept of coupling matrix synthesis has attracted recent researchers in the area of microwave filters. The most significant feature of the coupling matrix is the one-to-one correspondence between the values of the coupling matrix and physical components of the filter. Another advantage is the ability to reconfigure the coupling matrix through similarity transforms to arrive at a different coupling arrangement, corresponding to the available coupling elements of the particular microwave structure selected for the application. This can be done without going right back to the beginning of the network synthesis process to start again on a different network synthesis route, as would have to be done if the classical element extraction method was being used. In a word, the synthesis of generalized Chebyshev-coupled resonator filters refers to a process, relating the admittance matrix of n-coupled resonator network to generalized Chebyshev filtering function with pre-described transmission zeros at finite frequencies. Then, a coupling matrix corresponding to coupling elements in n-coupled resonator network is obtained, with a concomitant transformation of the coupling matrix to a physically realizable structure.

Coupling matrix for conventional filters like Butterworth and Chebyshev can be easily determined from the prototype values for a given specification. Band-pass filters (BPFs) considered for specific applications are assumed to be of a generalized Chebyshev type, as these filters are proven to meet the present-day demands like high out-of-band selectivity, in-band linearity, etc. The coupling matrix for these filters is obtained by first generating the polynomials representing the filter functions. The final realizable coupling matrix requires a series of similarity transforms to result in a matrix corresponding to a desired filter topology. The synthesis of coupling matrix for band-stop filters (BSF) follows very similar lines to fully canonical BPF with an exchange for the polynomials along with a constant representing the filter functions.

4.4 Lumped Element Filters

Lumped element filters consist only of ideal inductors, capacitors, and if necessary, transformers. Consider a low-pass filter (LPF) with a cutoff frequency of ω = 1 rad, operating between 1 Ω source and load resistors. The ideal amplitude response would require a network with an infinite number of elements. There are various approximations to this response which can be made and which only require a finite number of elements for their realization. Two most commonly used approximations to the ideal low-pass response are: (i) Butterworth or maximally flat approximation and (ii) Chebyshev or equiripple approximation.

Butterworth Approximation

$$L\,(\mathrm{dB}) = 20\log_{10}|S_{21}| = 10\log_{10}\left[1+\omega^{2n}\right] \tag{4.1}$$

Chebyshev Approximation

$$L\,(\mathrm{dB}) = 10\log_{10}\left[1+\left(10^{L_a/10}-1\right)\cos^2\left(n\cos^{-1}\omega'\right)\right]\mathrm{dB},\quad \omega\le 1 \tag{4.2}$$

$$L\,(\mathrm{dB}) = 10\log_{10}\left[1+\left(10^{L_a/10}-1\right)\cosh^2\left(n\cos^{-1}\omega'\right)\right]\mathrm{dB},\quad \omega\ge 1$$

Both these responses can be obtained by a ladder network. The total number of elements required in the ladder network is given by the value of n in (4.1) and (4.2). Increasing the value of n increases the filter's selectivity, but specifications which demand very high selectivity are best satisfied by using an elliptic function filter or a generalized filter. It should be noted that for all these filters, the closer the amplitude response approaches the ideal response, the more the group delay departs from being constant, particularly at the edges (Figure 4.3).

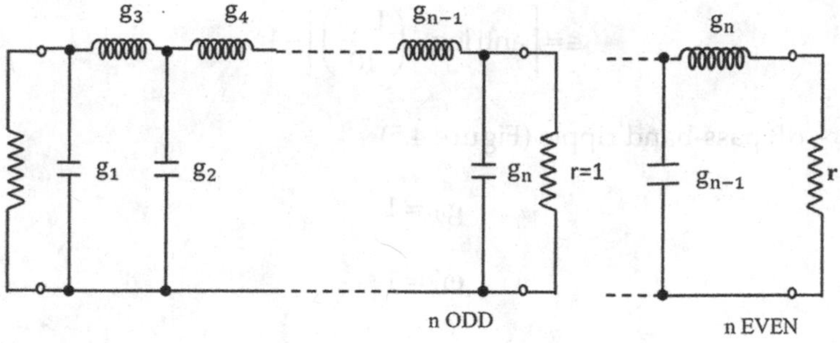

FIGURE 4.3
Prototype LPF and its design equations for maximally flat and Chebyshev response.

4.4.1 Maximally Flat Response

r = 1 for all n (Figure 4.4)

$$g_k = 2\sin\left[\frac{(2k-1)\pi}{2n}\right], \quad k = 1,2,\ldots n$$

$$A = 10\log_{10}\left(1+\omega'^{2n}\right)\text{dB}$$

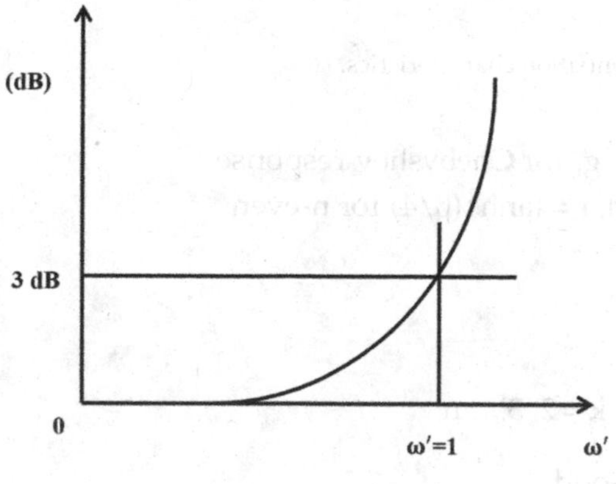

FIGURE 4.4
Maximally flat response.

4.4.2 Chebyshev LPF (Equal Ripple)

$$L_A(\omega') = 10\log_{10}\left\{1+\epsilon\cos^2\left[n\cos^{-1}\left(\frac{\omega'}{\omega'_1}\right)\right]\right\} \quad \omega' \le \omega'_1$$

$$L_A(\omega') = 10\log_{10}\left\{1+\epsilon\cosh^2\left[n\cosh^{-1}\left(\frac{\omega'}{\omega'_1}\right)\right]\right\} \quad \omega' \ge \omega'_1$$

$$\in = \left[\text{anti} \log_{10} \left(\frac{L_{Ar}}{10} \right) \right] - 1$$

For L_{Ar} – dB pass-band ripple (Figure 4.5)

$$g_0 = 1$$

$$\omega_1' = 1$$

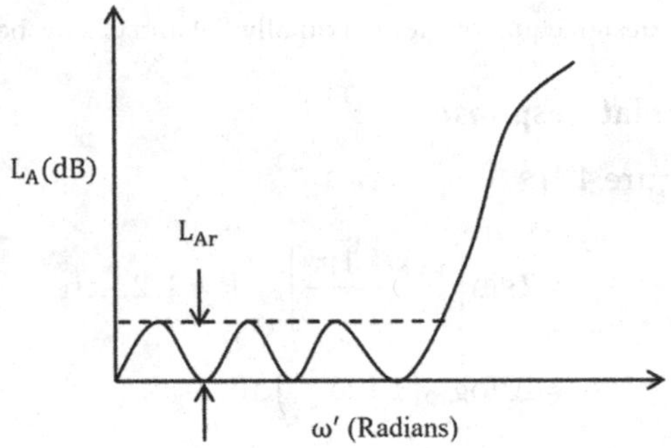

FIGURE 4.5
Chebyshev LPF attenuation characteristics.

Element value g_k for Chebyshev response
$r = 1$ for n-odd, $r = \tanh^2(\beta/4)$ for n-even

$$g_0 = 1$$

$$g_1 = \frac{2a_1}{\gamma}$$

$$g_k = \frac{4a_{k-1}a_k}{b_{k-1}g_{k-1}}, \; k = 2, 3, \ldots n$$

$g_{n+1} = 1$, for n-odd

$$g_{n+1} = \coth^2\left(\frac{\beta}{4}\right), \text{ for n-even}$$

$$\text{where } a_k = \sin\left(\frac{(2k-1)\pi}{2n}\right), \; k = 1, 2, \ldots n$$

$$b_k = \gamma^2 + \sin^2\left(\frac{k\pi}{n}\right), \; k = 1, 2, \ldots n$$

$$\beta = \ln\left(\coth\frac{L_{Ar}}{17.37}\right), \; L_{Ar} \text{ in dB}$$

$$\gamma = \sinh\left(\frac{\beta}{2n}\right)$$

$$L_A = 10\log_{10}\left[1+\left(10^{L_{Ar}/10}-1\right)\cos^2\left(n\cos^{-1}\omega'\right)\right]\text{dB}, \omega' \le 1$$

$$L_A = 10\log_{10}\left[1+\left(10^{L_{Ar}/10}-1\right)\cosh^2\left(n\cos^{-1}\omega'\right)\right]\text{dB}, \omega' \ge 1$$

Maximally Flat

Pass-band edge - The frequency ω_1', where the attenuation is (L_{Ar})

$$L_A(\omega') = 10\log_{10}\left\{1+\in\left[\left(\frac{\omega'}{\omega_1'}\right)^{2n}\right]\right\} \quad \omega' \ge \omega_1'$$

$$\in = \left[\text{anti}\log_{10}\left(\frac{L_{Ar}}{10}\right)\right] - 1$$

$$L_{Ar} = 3\text{ dB}$$

4.4.3 Butterworth Low-Pass Prototype

The element values g_k (Figure 4.6)

$$g_0 = 1$$

$$g_k = 2\sin\left[\frac{(2k-1)\pi}{2n}\right], \quad k = 1,2,\ldots n$$

$$g_{n+1} = 1 \quad \text{for all n}$$

$$g_{k|k=1\text{ to }n} = \begin{cases}\text{The inductance of a series coil (or)}\\ \text{The capacitance of a shuntcapacitor}\end{cases}$$

$$g_0 = \begin{cases}\text{The generator resistance } R_0' \; g_1 = C_1'\text{, but is}\\ \text{defined as the generator conductance } G_0' \text{ if } g_1 = L_1'\end{cases}$$

$$g_{n+1} = \begin{cases}\text{The loaded resistance } R_{n+1}' \text{ if } g_n = C_n'\text{, but is}\\ \text{defined as the load conductance } G_{n+1}' \text{ if } g_n = L_n'\end{cases}$$

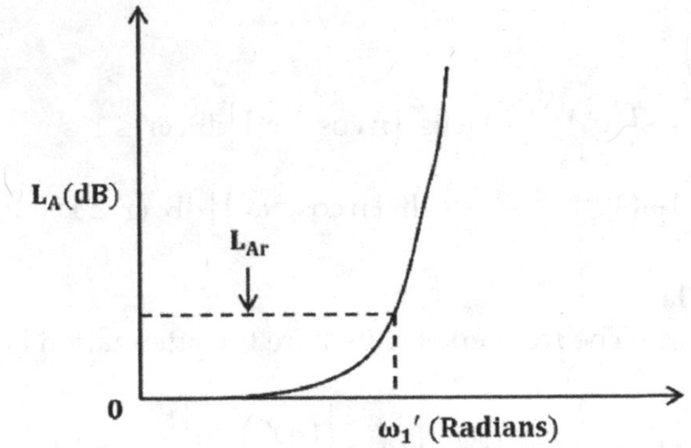

FIGURE 4.6
Maximally flat response.

4.5 Frequency and Impedance Scaling

If g_R – Resistance
  g_L – Inductors
  g_c – Inductors

New Prototype Values

$$R = R_0 g_R$$

$$L = \frac{R_0 g_L}{w_c}$$

$$C = \frac{g_c}{R_0 \omega_c}$$

Frequency Transformations

1. LPF to High-Pass Filter (HPF)

$$s' = \frac{\omega_0}{s}$$

2. LPF to BPF

$$s' = \frac{\omega_0}{BW}\left(\frac{s}{\omega_0} + \frac{\omega_0}{s}\right)$$

3. LPF to Band-Stop Filter

$$s' = \frac{BW}{\omega_0\left(\frac{s}{\omega_0} + \frac{\omega_0}{s}\right)}$$

where $BW = \omega_{c2} - \omega_{c1}$ and $\omega_0 = \omega_{c2}\omega_{c1}$

4.6 Band Pass Filter (BPF) Design Equations

- This allows determination of the odd and even characteristic line impedances,

$$Z_{Oo|i,i+1} = Z_O\left[1 - Z_O J_{i,i+1} + (Z_O J_{i,i+1})^2\right]$$

$$Z_{Oe|i,i+1} = Z_O\left[1 + Z_O J_{i,i+1} + (Z_O J_{i,i+1})^2\right]$$

- Indicates that i, i+1 refer to the overlapping elements, and Z_O is impedance at the ends of the filter structure.

$$Z_0 J_1 = \sqrt{\frac{\pi\Delta}{2g_1}}$$

$$Z_0 J_n = \sqrt{\frac{\pi\Delta}{2\sqrt{g_n g_{n+1}}}} \quad \text{for } n = 2,3,\ldots$$

$$Z_0 J_{n+1} = \sqrt{\frac{\pi\Delta}{2g_n g_{n+1}}} \quad \text{for } n = 1,2,3,\ldots$$

$$\Delta = \omega_2 - \omega_1 \text{ fractional bandwidth}$$

Even mode and Odd mode characteristic impedances Z_{Oe} and Z_{Oo} are computed using the above respective equations. Using the Nomograms the width 'w' amd spacing 's' of each section of guide quarter wavelength of each section of an edge coupled filter are found manually or through programming. Initially the order of the band pass filter is found from the specifications.

4.7 Filter Transformation

4.7.1 LPF to HPF Transformation

$$\text{For LPF} \quad S' = \frac{\omega_0}{s} \tag{4.3}$$

where
  s' = Prototype
  s = Actual frequency

Series Inductor:

$$\text{Impedance } (Z) = s'L_L \text{ (LPF)} \tag{4.4}$$

Substituting Equation (4.3) in (4.4) we get

$$Z = \left(\frac{\omega_0}{s}\right)L_L = \frac{1}{sC_H}$$

$$\boxed{Z = \frac{1}{sC_H}\text{(HPF)}} \tag{4.5}$$

where $c_H = \dfrac{1}{\omega_0 L_L}$

Shunt Capacitance:

$$\text{Admittance } (Y) = s'c_L \text{(LPF)} \tag{4.6}$$

Substituting Equation (4.3) in (4.6) we get

$$Y = \left(\frac{\omega_0}{s}\right)c_L = \frac{1}{sL_H}$$

$$\boxed{Y = \frac{1}{sL_H}\text{(HPF)}} \tag{4.7}$$

where $L_H = \dfrac{1}{\omega_0 c_L}$ (Figure 4.7)

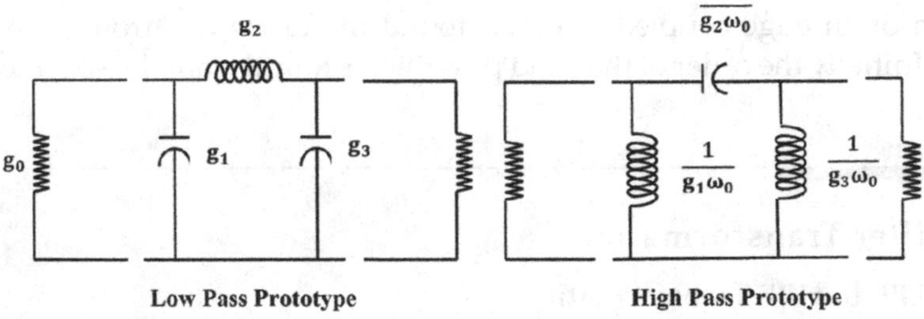

FIGURE 4.7
LPF to BPF transformation.

$$\text{For BPF } S' = \frac{\omega_0}{BW}\left(\frac{s}{\omega_0} + \frac{\omega_0}{s}\right) \tag{4.8}$$

Series Elements:

$$Z = g_2 s' \tag{4.9}$$

Substituting Equation (4.8) in (4.9) we get

$$Z = g_2\left(\frac{\omega_0}{BW}\left(\frac{s}{\omega_0} + \frac{\omega_0}{s}\right)\right)$$

$$Z = \frac{g_2 s}{BW} + \frac{g_2 \omega_0^{\ 2}}{s\,BW}$$

$$\boxed{Z = Z_1 + Z_2} \quad (\therefore \text{Series LC})$$

where $Z_1 = \dfrac{g_2}{BW}$ and $Z_2 = \dfrac{BW}{g_2 \omega_0^{\ 2}}$ (Figure 4.8)

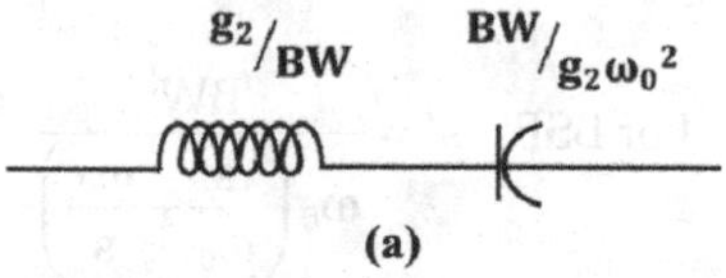

FIGURE 4.8A
Series branch transformation.

Therefore, the series becomes series inductor capacitor (LC).

Shunt Element:

$$Y = g_1 s' \tag{4.10}$$

Substituting Equation (4.8) in (4.10) we get

$$Y = g_1\left(\frac{\omega_0}{BW}\left(\frac{s}{\omega_0} + \frac{\omega_0}{s}\right)\right)$$

$$Y = \frac{g_1 s}{BW} + \frac{g_1 \omega_0{}^2}{s\,BW}$$

$$\boxed{Y = Y_1 + Y_2}$$

where $Y_1 = \frac{g_1 s}{BW}$ and $Y_2 = \frac{g\omega_0{}^2}{s BW}$

Therefore, parallel becomes parallel LC.

Transformed Circuit (Figure 4.8b):

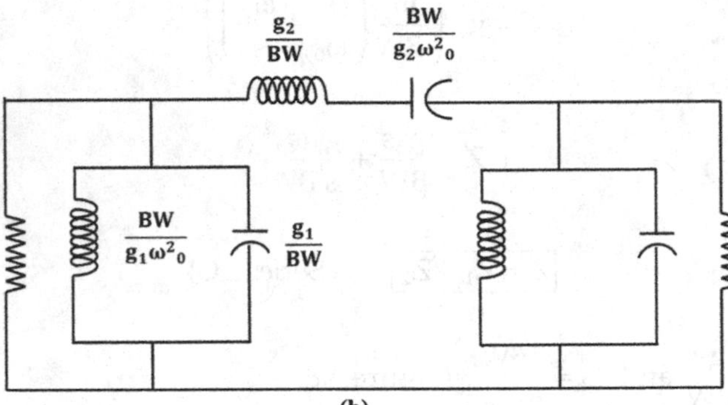

FIGURE 4.8B
LPF to BPF transformation.

$$\text{For BSF} \quad s' = \frac{BW}{\omega_0\left(\frac{s}{\omega_0} + \frac{\omega_0}{s}\right)} \tag{4.11}$$

Series Element:

$$Z = s' g_2 \tag{4.12}$$

Substituting Equation (4.11) in (4.12) we get

$$Z = \frac{g_2}{\frac{\omega_0}{BW}\left(\frac{s}{\omega_0} + \frac{\omega_0}{s}\right)}$$

$$Z = \frac{1}{\frac{s}{g_{2\,BW}} + \frac{\omega_0{}^2}{s\,g_{2\,BW}}}$$

$$\boxed{Z = \frac{1}{Y_1 + Y_2}}$$

where $Y_1 = \frac{s}{g_{2\,BW}}$ and $Y_2 = \frac{\omega_0^{\,2}}{s\,g_{2\,BW}}$ (Figure 4.9)

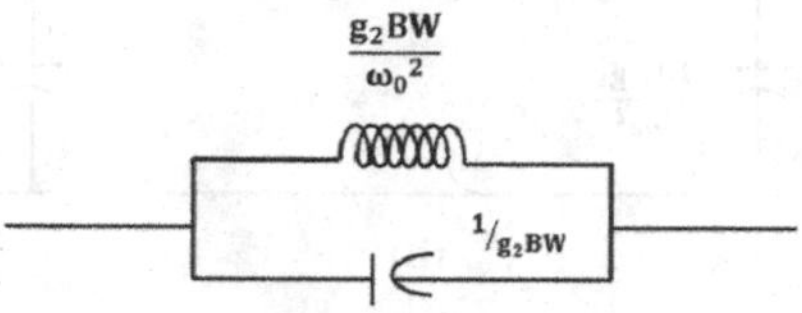

FIGURE 4.9A
Series branch transformation.

Therefore, the series becomes shunt LC.

Shunt Element:

$$Y = s' g_1 \tag{4.13}$$

Substituting Equation (4.11) in (4.13) we get

$$Y = \frac{g_1}{\frac{\omega_0}{BW}\left(\frac{s}{\omega_0} + \frac{\omega_0}{s}\right)}$$

$$Y = \frac{1}{\frac{s}{g_{1\,BW}} + \frac{\omega_0^{\,2}}{g_{1\,s\,BW}}}$$

$$\boxed{Y = \frac{1}{Z_1 + Z_2}}$$

where $Z_1 = \frac{s}{g_{1\,BW}}$ and $Z_2 = \frac{\omega_0^{\,2}}{g_1\,s\,BW}$

Therefore, the shunt element becomes series LC. It is the opposite of BPF.

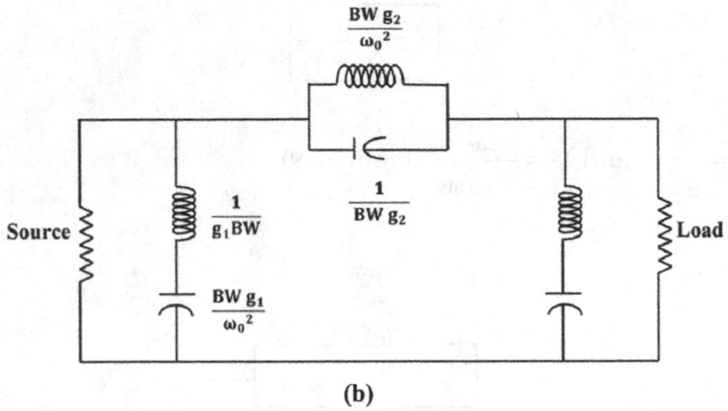

FIGURE 4.9B
Transformed circuit of BSF.

4.8 Filter Problems

1. The Butterworth LPF prototype: the element values g_k can be obtained from the following parameters. The element value $g_0 = 1$ and $g_k = 2\sin\left[\frac{(2k-1)\pi}{2n}\right]$; k = 1, 2, 3, … n.

Solution:
Given:

$$g_0 = 1$$

$$g_k = 2\sin\left[\frac{(2k-1)\pi}{2n}\right]; \quad k = 1,2,3,\ldots n.$$

For n = 1

$$g_0 = 1$$

$$g_1 = 2\sin\left[\frac{(2k-1)\pi}{2n}\right] = 2\sin\left[\frac{\pi}{2}\right] = 2$$

Therefore, $g_0 = 1$ and $g_1 = 2$
For n = 3

$$g_0 = 1$$

$$g_1 = 2\sin\left[\frac{(2-1)\pi}{6}\right] = 2\sin\left[\frac{\pi}{6}\right] = 1$$

$$g_2 = 2\sin\left[\frac{(4-1)\pi}{6}\right] = 2\sin\left[\frac{3\pi}{6}\right] = 2$$

$$g_3 = 2\sin\left[\frac{(6-1)\pi}{6}\right] = 2\sin\left[\frac{5\pi}{6}\right] = 1$$

For n = 5 (Figure 4.10)

$$g_0 = 1$$

$$g_1 = 2\sin\left[\frac{(2-1)\pi}{10}\right] = 2\sin\left[\frac{\pi}{10}\right] = 0.6180$$

$$g_2 = 2\sin\left[\frac{(4-1)\pi}{10}\right] = 2\sin\left[\frac{3\pi}{10}\right] = 1.6180$$

$$g_3 = 2\sin\left[\frac{(6-1)\pi}{10}\right] = 2\sin\left[\frac{5\pi}{10}\right] = 2$$

$$g_4 = 2\sin\left[\frac{(4-1)\pi}{10}\right] = 2\sin\left[\frac{7\pi}{10}\right] = 1.6180$$

$$g_5 = 2\sin\left[\frac{(6-1)\pi}{10}\right] = 2\sin\left[\frac{9\pi}{10}\right] = 0.6180$$

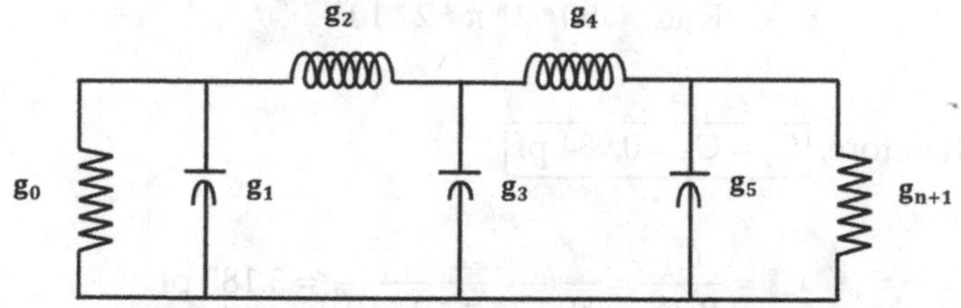

FIGURE 4.10
Five-section low-pass prototype filter with source and load impedences connected.

2. Design a maximally flax LPF with a cutoff frequency of 2 GHz, an impedance of 20 Ω, and at least 15-dB attenuation at 3 GHz. Compute elements of the LPF and draw the layout if n = 5.

Solution:
Given:

$$n = 5$$

Since (Figure 4.11)

$$g_1 = 0.6180 = C_1$$

$$g_2 = 1.6180 = L_1$$

$$g_3 = 2 = C_2$$

$$g_4 = 1.6180 = L_2$$

$$g_5 = 0.6180 = C_3$$

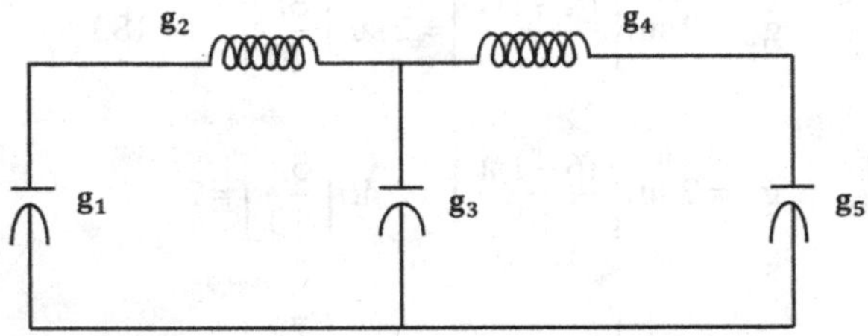

FIGURE 4.11
Five-section low-pass prototype filter.

$$C_1' = \frac{C_1}{R_0\omega_c} = \frac{0.6180}{50 * 2 * \pi * 2 * 10^9} = 0.984 \text{ pf}$$

Therefore, $\boxed{C_1' = C_3' = 0.984 \text{ pf}}$

$$C_2' = \frac{C_2}{R_0\omega_c} = \frac{2}{50 * 2 * \pi * 2 * 10^9} = 3.185 \text{ pf}$$

$$\boxed{C_2' = 3.185 \text{ pf}}$$

$$L_1' = \frac{R_0 L_1}{\omega_c} = \frac{50 * 1.618}{2 \times \pi \times 2 \times 10^9} = 6.44 \text{ nH}$$

Therefore, (Figures 4.12-4.15)

$$\boxed{L_1' = L_2' = 6.444 \text{ nH}}$$

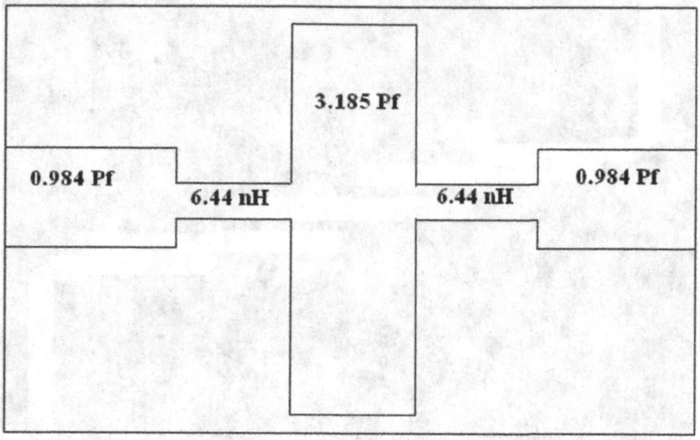

FIGURE 4.12
Layout of the LPF.

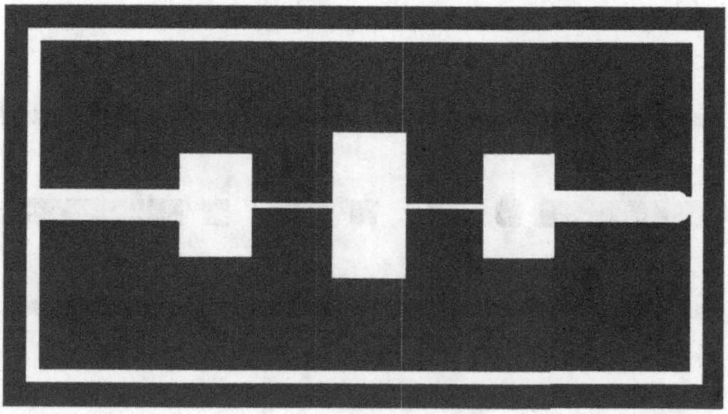

FIGURE 4.13
Low Pass Filter

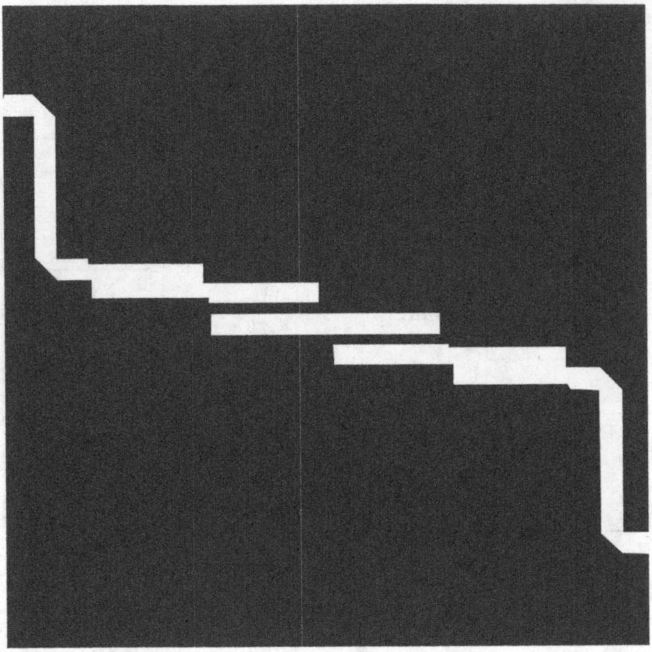

FIGURE 4.14
Edge Coupled Filter

FIGURE 4.15
End Coupled Filter

Bibliography

George L. Matthaei, Leo Young and E. M. T. Jones (1964). *Microwave Filters, Impedance-Matching Networks and Coupling Structures*, McGraw Hill, New York.

5

Microwave Amplifiers

5.1 Stability

Before designing a microwave amplifier or an oscillator, the important consideration to be taken is stability. The Rollet stability factor K and the delta factor Δ that are obtained from the scattering parameters of the transistor at a particular frequency facilitate the complete determination of analysis. Hence, it is the determining factor in both the amplifier and oscillator designs. There are systematic methods to determine the various kinds of stability for the given S-parameters of the transistor. The following are the methodology of stability, its need, and its applications. The determination of K and Δ can make one decide the suitability for an amplifier or oscillator.

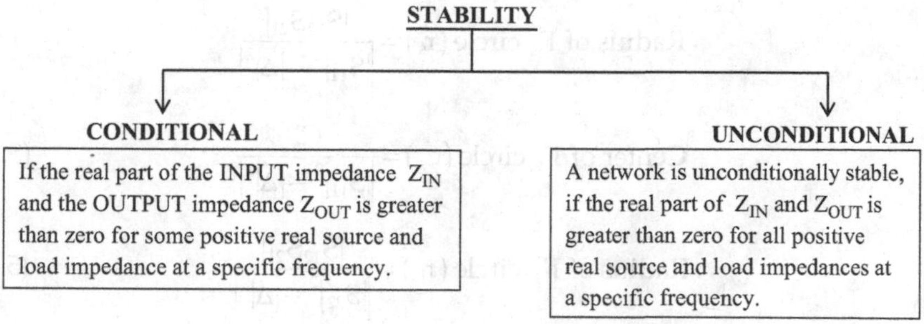

STABILITY
CONDITIONAL
If the real part of the INPUT impedance Z_{IN} and the OUTPUT impedance Z_{OUT} is greater than zero for some positive real source and load impedance at a specific frequency.
UNCONDITIONAL
A network is unconditionally stable, if the real part of Z_{IN} and Z_{OUT} is greater than zero for all positive real source and load impedances at a specific frequency.

Positive real source and load impedance means $|\Gamma_s| \leq 1$ and $|\Gamma_L| \leq 1$, where $\Gamma_s = \Gamma_{in}^*$ and $\Gamma_L = \Gamma_{out}^*$.

For unconditional stability, S_{11}, S_{22}, Γ_{in}, and Γ_{out} must be smaller than unity and the transistor inherent stability factor "K" must be greater than unity and positive "K."

$$K = \frac{1 + |\Delta|^2 - |S_{11}|^2 - |S_{22}|^2}{2|S_{12}S_{21}|} > 1 \tag{5.1}$$

where

$$\Delta = |S_{11}S_{22} - S_{12}S_{21}|^2 < 1 \quad (5.2)$$

$$\text{Load Reflection Coefficient } (\Gamma_L) = \frac{Z_L - Z_0}{Z_L + Z_0} \quad (5.3)$$

$$\text{Source Reflection Coefficient } (\Gamma_s) = \frac{Z_s - Z_0}{Z_s + Z_0} \quad (5.4)$$

$$\text{Input Reflection Coefficient } (\Gamma_{in}) = S_{11} + \frac{S_{12}S_{21}\Gamma_L}{1 - S_{22}\Gamma_L} = \frac{S_{11} - \Delta\Gamma_L}{1 - S_{22}\Gamma_L} \quad (5.5)$$

$$\text{Output Reflection Coefficient } (\Gamma_{out}) = S_{22} + \frac{S_{12}S_{21}\Gamma_s}{1 - S_{11}\Gamma_s} = \frac{S_{22} - \Delta\Gamma_s}{1 - S_{11}\Gamma_s} \quad (5.6)$$

The boundary conditions for stability are given by

$$|\Gamma_{in}| = \left| S_{11} + \frac{S_{12}S_{21}\Gamma_L}{1 - S_{22}\Gamma_L} \right| = 1 \quad (5.7)$$

$$|\Gamma_{out}| = \left| S_{22} + \frac{S_{12}S_{21}\Gamma_s}{1 - S_{11}\Gamma_s} \right| = 1 \quad (5.8)$$

$$\text{Radius of } \Gamma_s \text{ circle } (r_s) = \frac{|S_{12}S_{21}|}{\left| |S_{11}|^2 - |\Delta|^2 \right|} \quad (5.9)$$

$$\text{Center of } \Gamma_s \text{ circle } (c_s) = \frac{C_s^*}{\left| |S_{11}|^2 - |\Delta|^2 \right|} \quad (5.10)$$

$$\text{Radius of } \Gamma_L \text{ circle } (r_L) = \frac{|S_{12}S_{21}|}{\left| |S_{22}|^2 - |\Delta|^2 \right|} \quad (5.11)$$

$$\text{Center of } \Gamma_L \text{circle } (c_L) = \frac{C_L^*}{\left| |S_{22}|^2 - |\Delta|^2 \right|} \quad (5.12)$$

where

$\Delta = S_{11}S_{22} - S_{12}S_{21}$

$C_s = S_{11} - \Delta S_{22}^*$

$C_L = S_{22} - \Delta S_{11}^*$

For unconditional stability, any passive source and load impedances in a two-port network must produce the stability circles completely out of the

Smith chart. If the stability circles are overlapped with the Smith chart, the stability is still conditional (Figure 5.1).

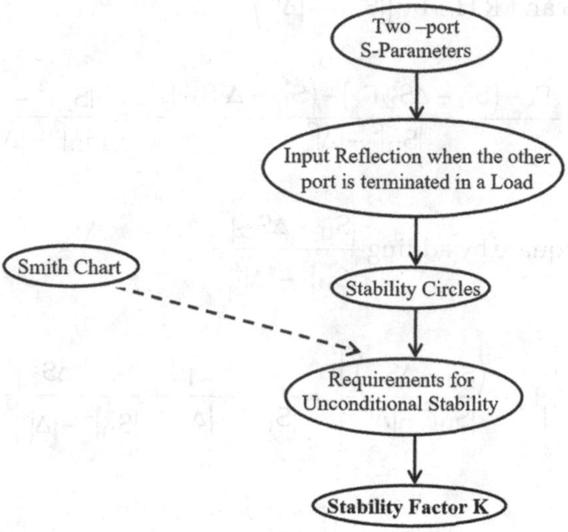

FIGURE 5.1
Flow concepts needed for evaluation of the stability factor K.

5.2 Input and Output Stability Circles

The boundary conditions for output stability are given by

$$|\Gamma_{out}| = \left| S_{22} + \frac{S_{12}S_{21}\Gamma_s}{1 - S_{11}\Gamma_s} \right| = 1 \tag{5.13}$$

$$\Delta = S_{11}S_{22} - S_{12}S_{21}$$

$$\therefore |S_{22}(1 - S_{11}\Gamma_s) + S_{12}S_{21}| = |1 - S_{11}\Gamma_s| \tag{5.14}$$

$$|S_{22} - \Delta\Gamma_s| = |1 - S_{11}\Gamma_s| \tag{5.15}$$

Squaring on both sides,

$$|S_{22}|^2 + |\Delta|^2|\Gamma_s|^2 - (\Delta\Gamma_s S_{22}^* + \Delta^*\Gamma_s^* S_{22}) = 1 + |S_{11}|^2|\Gamma_s|^2 - (S_{11}^*\Gamma_s^* + S_{11}\Gamma_s) \tag{5.16}$$

$$|S_{22}|^2 + |\Delta|^2\Gamma_s\Gamma_s^* - (\Delta\Gamma_s S_{22}^* + \Delta^*\Gamma_s^* S_{22}) = 1 + |S_{11}|^2\Gamma_s\Gamma_s^* - (S_{11}^*\Gamma_s^* + S_{11}\Gamma_s) \tag{5.17}$$

Taking R.H.S to L.H.S and all except $|S_{22}|^2$ to R.H.S,

$$\left(|S_{11}|^2 - |\Delta|^2\right)\Gamma_s\Gamma_s^* - \left(S_{11} - \Delta S_{22}^*\Gamma_s\right) + \left(S_{11}^* - \Delta^* S_{22}\right)\Gamma_s^* = |S_{22}|^2 - 1 \quad (5.18)$$

Dividing L.H.S and R.H.S by $\left(|S_{11}|^2 - |\Delta|^2\right)$,

$$\frac{\Gamma_s\Gamma_s^* - \left(S_{11} - \Delta S_{22}^*\Gamma_s\right) - \left(S_{11}^* - \Delta^* S_{22}\right)\Gamma_s^*}{|S_{11}|^2 - |\Delta|^2} = \frac{|S_{22}|^2 - 1}{|S_{11}|^2 - |\Delta|^2} \quad (5.19)$$

Complete the square by adding $\dfrac{\left|S_{11} - \Delta S_{22}^*\right|^2}{\left|\,|S_{11}|^2 - |\Delta|^2\right|^2}$

$$\left|\Gamma_s - \frac{\left(S_{11} - \Delta S_{22}^*\right)^*}{|S_{11}|^2 - |\Delta|^2}\right|^2 = \frac{|S_{22}|^2 - 1}{|S_{11}|^2 - |\Delta|^2} + \frac{\left|S_{11} - \Delta S_{22}^*\right|^2}{\left|\,|S_{11}|^2 - |\Delta|^2\right|^2}$$

or

$$\left|\Gamma_s - \frac{\left(S_{11} - \Delta S_{22}^*\right)}{|S_{11}|^2 - |\Delta|^2}\right| = \frac{|S_{12}S_{21}|}{|S_{11}|^2 - |\Delta|^2} \quad (5.20)$$

where

Center $C_s = \dfrac{C_1^*}{|S_{11}|^2 - |\Delta|^2}$ and radius $r_s = \dfrac{|S_{12}S_{21}|}{|S_{11}|^2 - |\Delta|^2}$ on the Γ_s plane

$C_1 = S_{11} - \Delta S_{22}^*$

Similarly, for output stability circle on the Γ_L plane

$$C_s = \frac{C_2^*}{|S_{22}|^2 - |\Delta|^2} \quad (5.21)$$

$$r_s = \frac{|S_{12}S_{21}|}{|S_{22}|^2 - |\Delta|^2} \quad (5.22)$$

$$C_2 = S_{22} - \Delta S_{11}^*$$

The stability circle defines the boundary between the stable and unstable regions on the Smith chart.

- This means that either the inside or the outside of the circle can represent the stable region.

- The most suitable point is the center of the Smith chart, that is, the point $\Gamma_L = 0$. At this point, Equation (5.7) reduces to $\Gamma_{in} = S_{11}$. Recalling the conditions for stability gives $|S_{11}| < 1$. If this condition is satisfied, then the Smith chart center represents a stable region.

The two possible cases of this condition are illustrated in Figure 5.2a,b. The first condition is where the stability circle encloses the origin and hence its interior represents the stable region. The second condition illustrates the case where the origin is outside the circle and the exterior of the circle represents the stable region.

A schematic diagram of the four different regions of stability circles is provided in Figure 5.2.

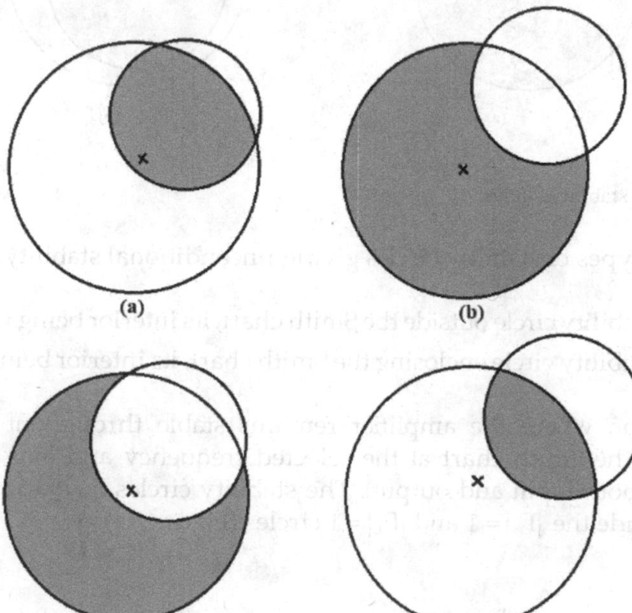

FIGURE 5.2
Stability circles. (a and b) The Smith chart center is stable. (c and d) The Smith chart center is unstable.

If the stability conditions are not satisfied for the origin at $\Gamma_L = 0$, then $|S_{11}| < 1$ becomes $|S_{11}| > 1$, and the origin represents an unstable point. The two cases of this condition are shown in Figure 5.2c,d. Figure 5.2c is the case where the stability circle encircles the origin and the exterior is the stable region. The case where the circle does not encircle the origin, the interior represents the stable region (shown in Figure 5.2d).

5.3 Unconditional Stability

The two types of stability circles illustrating a stable region on the inside and outside are shown in Figure 5.3a,b.

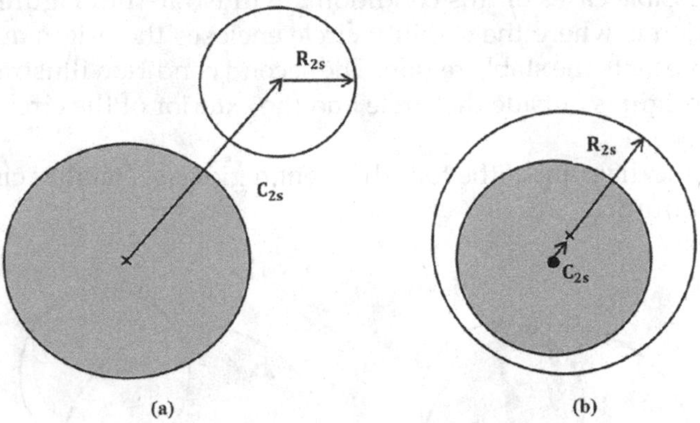

FIGURE 5.3
Unconditional stability circles.

The two types of stability circles giving unconditional stability:

a. The stability circle outside the Smith chart, its interior being unstable.
b. The stability circle enclosing the Smith chart, its interior being stable.

The situation where the amplifier remains stable throughout the entire domain of the Smith chart at the selected frequency and bias conditions (applies to both input and output). The stability circles have to reside completely outside the $|\Gamma_s| = 1$ and $|\Gamma_L| = 1$ circles (Figure 5.4).

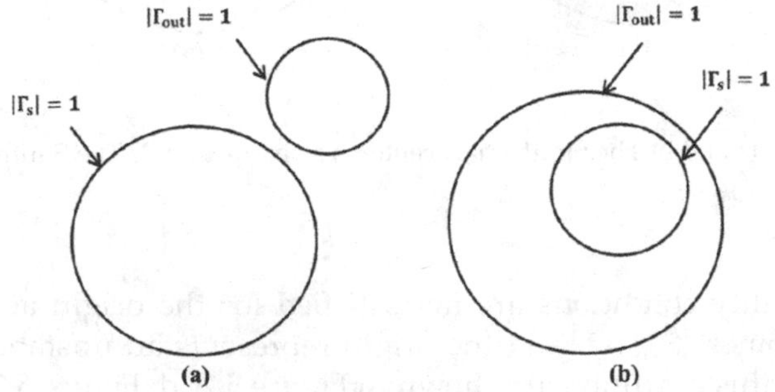

FIGURE 5.4
Unconditional stability in the Γ_s and Γ_{out} planes for $|S_{11}| < 1$. (a) $|\Gamma_{out}| = 1$ circles must reside outside. (b) $|\Gamma_s| = 1$ circles must reside inside.

$$|\Gamma_{in}| = \left|\frac{S_{11} - \Gamma_L \Delta}{1 - S_{22}\Gamma_L}\right| < 1 \tag{5.23}$$

$$|\Gamma_{out}| = \left|\frac{S_{22} - \Gamma_s \Delta}{1 - S_{11}\Gamma_s}\right| < 1 \tag{5.24}$$

where $\Delta = S_{11}S_{22} - S_{12}S_{21}$.

If $\Gamma_L = 0$, then $|\Gamma_{in}| = |S_{11}|$ (two cases $|S_{11}| < 1$ & $|S_{11}| > 1$). For $|S_{11}| < 1$, the origin (the point $\Gamma_L = 0$) is part of the stable region (Figure 5.5).

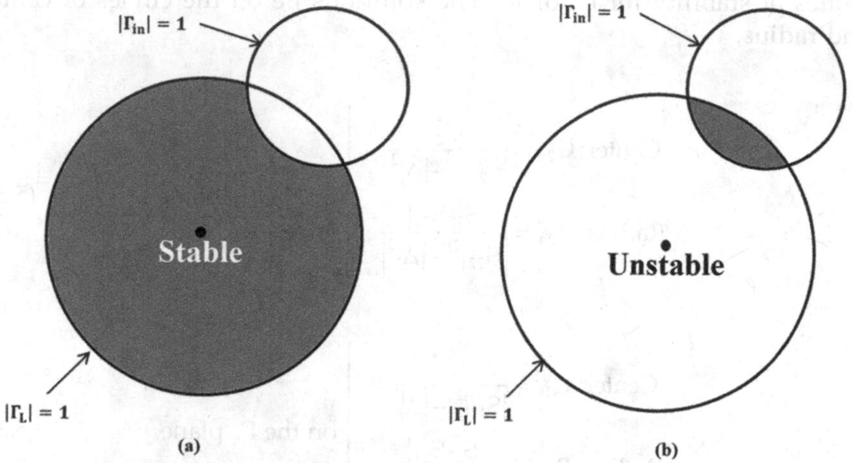

FIGURE 5.5
Output stability circles denoting stable and unstable regions. (a) Shaded region is stable, since $|S_{11}| < 1$. (b) Shaded region excludes the origin $\Gamma_L = 0$, since $|S_{11}| > 1$.

5.4 Stability Circles

$$\text{Input Reflection Coefficient } (\Gamma_{in}) = S_{11} + \frac{S_{12}S_{21}\Gamma_L}{1 - S_{22}\Gamma_L} \tag{5.25}$$

$$\text{Output Reflection Coefficient } (\Gamma_{out}) = S_{22} + \frac{S_{12}S_{21}\Gamma_s}{1 - S_{11}\Gamma_s} \tag{5.26}$$

For stability, the boundary conditions are

$$|\Gamma_{in}| < 1 \tag{5.27}$$

$$|\Gamma_{out}| < 1 \tag{5.28}$$

Equations (5.25) and (5.26) are standard bilinear under each transformation, a circle of constant Γ_L or Γ_s is transformed into a circle in the Γ_{in} or Γ_{out} plane. This is a significant result since it is now possible to define stability region on the Smith chart representing the Γ_L or Γ_s plane.

Setting

$$|\Gamma_{in}| = 1 \tag{5.29}$$

$$|\Gamma_{out}| = 1 \tag{5.30}$$

Substituting these Equations in (5.25) and (5.26) will give the limiting values of stability for Γ_L or Γ_s. The solutions lie on the circles of center and radius.

$$\left.\begin{aligned} \text{Center: } C_{2s} &= \frac{C_2^*}{|S_{22}|^2 - |\Delta|^2} \\ \text{Radius: } R_{2s} &= \frac{|S_{12}S_{21}|}{\left||S_{22}|^2 - |\Delta|^2\right|} \end{aligned}\right\} \text{on the } \Gamma_L \text{ plane} \tag{5.31}$$

$$\left.\begin{aligned} \text{Center: } C_{1s} &= \frac{C_1^*}{|S_{11}|^2 - |\Delta|^2} \\ \text{Radius: } R_{1s} &= \frac{|S_{12}S_{21}|}{\left||S_{11}|^2 - |\Delta|^2\right|} \end{aligned}\right\} \text{on the } \Gamma_s \text{ plane} \tag{5.32}$$

where

$C_1 = S_{11} - \Delta S_{22}^*$

$C_2 = S_{22} - \Delta S_{11}^*$

$\Delta = S_{11}S_{22} - S_{12}S_{21}$

The stability circle defines the boundary between the stable and unstable regions on the Smith chart.

- This means that either the inside or the outside of the circle can represent the stable region.
- The most suitable point is the center of the Smith chart, that is, the point $\Gamma_L = 0$. At this point, Equation (5.25) reduces to $\Gamma_{in} = S_{11}$. Recalling the conditions for stability gives $|S_{11}| < 1$. If this condition is satisfied, then the Smith chart center represents a stable region.

5.5 Input Stability Circles

If $|S_{22}| < 1$, then it leads to the conclusion that the center $(\Gamma_s = 0)$ must be stable; otherwise, the center becomes unstable for $|S_{22}| > 1$, which is shown in Figure 5.6.

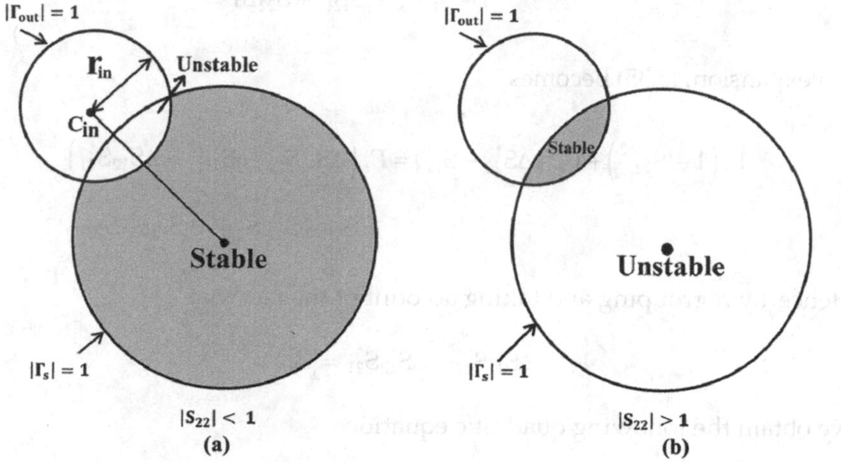

FIGURE 5.6
Input stability circles denoting (a) stable and (b) unstable regions.

5.6 Constant Gain Circles

Once the stability is desired from the stability circles, input constant gain circles and output constant gain circles are needed to obtain the required overall amplifier gain. The procedure to draw input and output constant gain circles are detailed as follows.

For matching $\Gamma_s = \Gamma_{in}^*$ and $\Gamma_L = \Gamma_{out}^*$

$$\therefore \Gamma_s = S_{11}^* + \frac{S_{12}^* S_{21}^* \Gamma_L^*}{1 - S_{22}^* \Gamma_L^*} \tag{5.33}$$

$$\Gamma_L^* = S_{22}^* + \frac{S_{12} S_{21} \Gamma_s}{1 - S_{11}\Gamma_s} = \frac{S_{22} - \Delta\Gamma_s}{1 - S_{11}\Gamma_s} \tag{5.34}$$

where $\Delta = S_{11}S_{22} - S_{12}S_{21}$.

Substituting Equation (5.34) in (5.33) for Γ_L^* in Γ_s

$$\Gamma_s = S_{11}^* + \frac{S_{12}^* S_{21}^* \left(\dfrac{S_{22} - \Delta\Gamma_s}{1 - S_{11}\Gamma_s} \right)}{1 - S_{22}^* \left(\dfrac{S_{22} - \Delta\Gamma_s}{1 - S_{11}\Gamma_s} \right)}$$

$$\Gamma_s = S_{11}^* + \frac{S_{12}^* S_{21}^* (S_{22} - \Delta\Gamma_s)}{1 - S_{11}\Gamma_s - |S_{22}|^2 + S_{22}^* \Delta\Gamma_s} \tag{5.35}$$

On expansion, (5.35) becomes

$$\Gamma_s\left(1 - |S_{22}|^2\right) + \Gamma_s{}^2\left(\Delta S_{22}^* - S_{11}\right) = \Gamma_s\left(\Delta S_{11}^* S_{22}^* - |S_{11}|^2 - \Delta S_{12}^* S_{21}^*\right)$$
$$+ S_{11}^* - S_{11}^* |S_{22}|^2 + S_{12}^* S_{21}^* S_{22}$$

Hence, by regrouping and taking account of the fact that

$$\Delta S_{11}^* S_{22}^* - \Delta S_{12}^* S_{21}^* = |\Delta|^2$$

we obtain the following quadratic equation

$$\left(S_{11} - \Delta S_{22}^*\right)\Gamma_s{}^2 + \left(|\Delta|^2 - |S_{11}|^2 + |S_{22}|^2 - 1\right)\Gamma_s + \left(S_{11}^* - S_{11}^*|S_{22}|^2 + S_{12}^* S_{21}^* S_{22}\right) = 0$$

The constant term is equal to C_1^* such that

$$C_1^* = S_{11}^* - \Delta^* S_{22} = \left(S_{11} - \Delta S_{22}^*\right)^* \tag{5.36}$$

$$\therefore \Gamma_s = \frac{B_1 \pm \sqrt{\left(B_1{}^2 - 4|\Delta C_1|^2\right)}}{2C_1} \tag{5.37}$$

where

$B_1 = 1 + |S_{11}|^2 - |S_{22}|^2 - |\Delta|^2$

$C_1 = S_{11} - \Delta S_{22}^*$

$\Delta = S_{11}S_{22} - S_{12}S_{21}$

Similarly,

$$\Gamma_s = \frac{B_2 \pm \sqrt{\left(B_2{}^2 - 4|\Delta C_2|^2\right)}}{2C_2}$$

where

$B_2 = 1 + |S_{22}|^2 - |S_{11}|^2 - |\Delta|^2$

$C_2 = S_{22} - \Delta S_{11}^*$

Calculation of $B_1{}^2 - 4|C_1|^2$

By expansion

$$B_1{}^2 = 1 + 2|S_{11}|^2 - 2|S_{22}|^2 - 2|\Delta|^2 + |S_{11}|^4 + |S_{22}|^4 + |\Delta|^4 - 2|S_{11}|^2|S_{22}|^2 - 2|\Delta|^2|S_{11}|^2 + 2|\Delta|^2|S_{22}|^2$$

$$|C_1|^2 = |S_{11}|^2 - \Delta^* S_{11}S_{22} - \Delta S_{11}^* S_{22}^* + |\Delta|^2|S_{22}|^2 = |S_{11}|^2 - 2|S_{11}|^2|S_{22}|^2 + |\Delta|^2|S_{22}|^2 + S_{12}^* S_{21}^* S_{11}S_{22} + S_{12}S_{21}S_{11}^* S_{22}^* \tag{5.38}$$

Now, by taking the product $|\Delta|^2 = \Delta\Delta^*$, we have

$$S_{12}^* S_{21}^* S_{11}S_{22} + S_{12}S_{21}S_{11}^* S_{22}^* = |S_{11}|^2|S_{22}|^2 + |S_{12}|^2|S_{21}|^2 - |\Delta|^2$$

Hence,

$$|C_1|^2 = |S_{11}|^2 - |S_{11}|^2|S_{22}|^2 + |S_{12}|^2|S_{21}|^2 - |\Delta|^2 + |\Delta|^2|S_{22}|^2$$

$$\therefore B_1{}^2 - 4|C_1|^2 = 1 - 2|S_{11}|^2 - 2|S_{22}|^2 + 2|\Delta|^2 + |S_{11}|^4 + |S_{22}|^4 + |\Delta|^4 + 2|S_{11}|^2|S_{22}|^2 - 2|\Delta|^2|S_{11}|^2 - 2|\Delta|^2|S_{22}|^2 - 4|S_{12}|^2|S_{12}|^2$$

This expression can finally be written in the form

$$B_1{}^2 - 4|C_1|^2 = 4\left(k^2 - 1\right)|S_{12}S_{12}|^2 \tag{5.39}$$

where $K = \dfrac{1 + |\Delta|^2 - |S_{11}|^2 - |S_{22}|^2}{2|S_{12}S_{21}|}$

The R.H.S is in variant; hence,

$$B_1{}^2 - 4|C_1|^2 = B_2{}^2 - 4|C_2|^2 \tag{5.40}$$

5.7 Design Procedure for Stability Circles

Check for inherent stability of active device using

$$K = \frac{1 + |\Delta|^2 - |S_{11}|^2 - |S_{22}|^2}{2|S_{12}S_{21}|} > 1 \tag{5.41}$$

$$|S_{11}| < 1 \text{ and } |S_{11}| > 1 \tag{5.42}$$

For any inherent stability, any load and source reflection coefficient within the Smith chart represents a stable termination.

For a potentially unstable device, the load stability is calculated and plotted on the Smith chart using the following equations:

$$\text{Center:} \quad C_{2s} = \frac{C_2^*}{|S_{22}|^2 - |\Delta|^2} \tag{5.43}$$

$$\text{Radius:} \quad R_{2s} = \frac{|S_{12}S_{21}|}{\left||S_{22}|^2 - |\Delta|^2\right|} \tag{5.44}$$

Similarly, the source stability circle is given in the following equations:

$$\text{Center:} \quad C_{1s} = \frac{C_1^*}{|S_{11}|^2 - |\Delta|^2} \tag{5.45}$$

$$\text{Radius:} \quad R_{1s} = \frac{|S_{12}S_{21}|}{\left||S_{11}|^2 - |\Delta|^2\right|} \tag{5.46}$$

The radius of the load stability circle is given by

$$Q_s = \frac{|S_{12}S_{21}|}{\left||S_{22}|^2 - |\Delta|^2\right|} \tag{5.47}$$

The load and source terminations selected, that is, Γ_L and Γ_s, must fall within the stable region as defined, normal classification of design such as, maximum gain design or low-noise design.

Broadband design adds additional constraints, but it is basically the same as the maximum gain or the low-noise amplifier design.

5.8 Noise Figure

Noise figure plays an important role in determining the quality of the amplifier. Here, the overall noise figure is considered rather than the input and output noise figures. System performance is determined by the noise figure.

Noise figure is defined as the ratio between the signal-to-noise ratio at the input and the signal-to-noise ratio at the output.

$$F = \frac{S_{in}/N_{in}}{S_o/N_o} = \frac{S_{in}N_o}{S_oN_{in}} \tag{5.48}$$

The noise figure of a two-port microwave amplifier is given by

$$F = F_{min} + \frac{r_n}{g_s}\left|Y_s - Y_0\right|^2 \tag{5.49}$$

$$F = F_{min} + \frac{r_n}{g_s}\left[\left(g_s - g_0\right)^2 + \left(b_s - b_0\right)^2\right] \tag{5.50}$$

where

F_{min} = Minimum noise figure, which is the function of the device operating frequency and current.

$r_n = \frac{R_n}{Z_0}$ is the normalized noise resistance of the two ports.

$Y_s = g_s + jb_s$ is the normalized source admittance.

$Y_0 = g_0 + jb_0$ is the normalized optimum source admittance which results in the minimum noise figure.

The normalized source admittance can be written in terms of the source reflection coefficient

$$Y_s = \frac{1 - \Gamma_s}{1 + \Gamma_s} \tag{5.51}$$

Similarly, the normalized optimum source admittance can be expressed as

$$Y_0 = \frac{1 - \Gamma_0}{1 + \Gamma_0} \tag{5.52}$$

where Γ_0 is the optimum source reflection coefficient, which results in the minimum noise figure.

$$F = F_{min} + \frac{4r_n|\Gamma_s - \Gamma_0|^2}{\left(1-|\Gamma_s|^2\right)|1+\Gamma_0|^2} \tag{5.53}$$

The resistance r_n can be found by noise figure (F) for $\Gamma_s = 0$ when a 50 Ω source resistance is used.

$$r_n = \left(F_{\Gamma_s=0} - F_{min}\right)\frac{|1+\Gamma_0|^2}{4|\Gamma_0|^2} \tag{5.54}$$

To determine the noise figure circle for a given noise figure (F_i), we define a noise figure parameter N_i.

$$N_i = \frac{|\Gamma_s - \Gamma_0|^2}{1-|\Gamma_s|^2} = \frac{F_i - F_{min}}{4r_n}\left(1-|\Gamma_s|^2\right) \tag{5.55}$$

Equation (5.55) can be written as

$$(\Gamma_s - \Gamma_0)\left(\Gamma_s^* - \Gamma_0^*\right) = N_i\left(1-|\Gamma_s|^2\right) = |\Gamma_s - \Gamma_0|^2$$

or

$$|\Gamma_s|^2(1+N_i) + |\Gamma_0|^2 - 2\,\mathrm{Re}\left(\Gamma_s\Gamma_0^*\right) = |\Gamma_s - \Gamma_0|^2 \tag{5.56}$$

Multiplying both sides by $(1 + N_i)$,

$$|\Gamma_s|^2(1+N_i) + |\Gamma_0|^2 - 2(1+N_i)\mathrm{Re}\left(\Gamma_s\Gamma_0^*\right) = N_i^{\,2} + N_i\left(1-|\Gamma_0|^2\right)$$

$$|\Gamma_s(1+N_i) - \Gamma_0|^2 = N_i^{\,2} + N_i\left(1-|\Gamma_0|^2\right)$$

Then,

$$\left|\Gamma_s - \frac{\Gamma_0}{1+N_i}\right|^2 = \frac{N_i^{\,2} + N_i\left(1-|\Gamma_0|^2\right)}{(1+N_i)^2} \tag{5.57}$$

This equation represents a family of circle in terms of N_i. The circle center and radius are given by

$$C_{F_i} = \frac{\Gamma_0}{1+N_i} \tag{5.58}$$

$$r_{F_i} = \frac{1}{1+N_i}\left[N_i^{\,2} + N_i\left(1-|\Gamma_0|^2\right)\right]^{\frac{1}{2}} \tag{5.59}$$

From this equation, $N_i = 0$ when $F_i = F_{min}$, and the center of the F_{min} circle is the zero radius is located at Γ_0 on the Smith chart. The centers of the other noise figure circles lie along the optimum source reflection coefficient Γ_0 vector. If a given source impedance is located along a specific noise circle, that impedance would result in a specific noise figure in decibels at that point.

5.9 Low-Noise Amplifier

A set of noise figure circles is generated using

$$\text{Center:} \quad C_{F_i} = \frac{\Gamma_{opt}}{1+N_i} \quad [\because \Gamma_0 = \Gamma_{opt}] \tag{5.60}$$

$$\text{Radius:} \quad r_{F_i} = \frac{1}{1+N_i}\sqrt{\left[N_i^{\,2} + N_i\left(1-|\Gamma_{opt}|^2\right)\right]} \tag{5.61}$$

Γ_s is selected to lie on the required noise figure circle, and Γ_L is calculated using

$$\Gamma_L = \left(\frac{S_{22} - \Delta\Gamma_s}{1 - S_{11}\Gamma_s}\right) \tag{5.62}$$

Both Γ_L and Γ_s must lie in the stable region. The transducer gain is checked using

$$G_T = \frac{|S_{21}|^2\left(1-|\Gamma_S|^2\right)\left(1-|\Gamma_L|^2\right)}{|1 - S_{11}\Gamma_s - S_{22}\Gamma_L + \Delta\Gamma_s\Gamma_L|} \tag{5.63}$$

If Γ_L does not lie in the stable region or the transducer gain is not high enough, the process of selecting Γ_s can be repeated until a suitable noise gain compromise is selected (Figure 5.7).

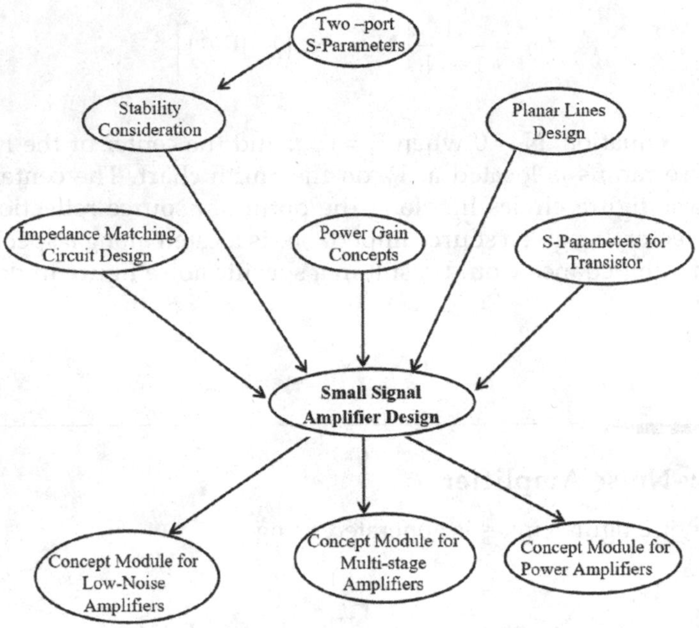

FIGURE 5.7
Concept module for small-signal microwave amplifier design. Incoming arrows show five prerequisite modules. The three outgoing arrows show possible concept module.

5.10 S-Parameters and Signal Flow Graphs

A signal flow graph is a pictorial representation of a system normally described by a set of simultaneous equations. In microwave circuit analysis, circuits are described in terms of traveling "power" waves, a_s and b_s, related to each other by S-parameters in the form of linear simultaneous equations. Hence, the signal flow graph technique can easily be adopted to represent linear microwave circuits pictorially via S-parameters, and furthermore, it can also be used to simplify circuits for analysis.

Let us first represent a two-port network with parameters S, that is,

$$b_1 = S_{11}a_1 + S_{21}a_2 \tag{5.64}$$

$$b_2 = S_{21}a_1 + S_{22}a_2 \tag{5.65}$$

By considering variables, both dependent (b_s) and independent (a_s), as nodes and the s_{ij}s as branches, (5.64) and (5.65) can be represented pictorially by a signal flow graph as shown in Figure 5.8.

Consider now a signal generator of source voltage V, and source impedance Z_s, as shown in Figure 5.9a. The power-delivering capacity of the generator can be described by P_{avso}, defined as the power available from the source (generator) to be delivered to a load equal to certain system impedance Z_0, that is,

$$P_{avso} = \left(\frac{V_s}{Z_s + Z_0} \right)^2 Z_0 \tag{5.66}$$

By defining a "power"-wave variable b_s equal to $\sqrt{P_{avso}}$, that is, defining b_s as the square root of the power available from the source to a load equal to the system impedance (chosen) Z_0, the generator can be represented by a signal flow graph diagram shown in Figure 5.9b.

In Figure 5.9b, Γ_s is the reflection coefficient of the source, defined with respect to the system impedance Z_0 as

$$\Gamma_s = \frac{Z_s - Z_0}{Z_s + Z_0} \tag{5.67}$$

The signal flow diagram for a load Z_L can be similarly deduced by defining $\Gamma_L = (Z_L - Z_0)/(Z_L + Z_0)$ as shown in Figure 5.10.

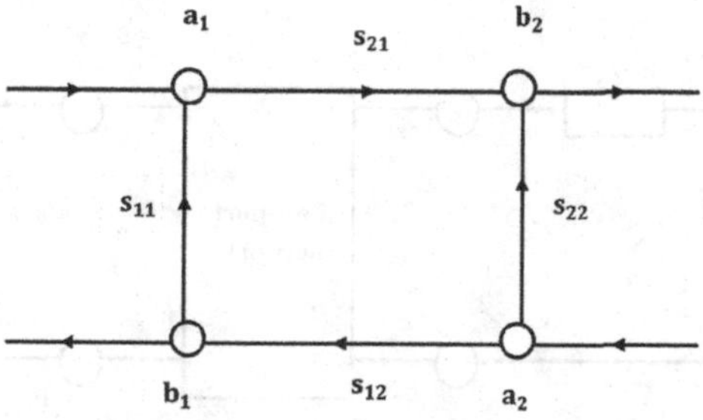

FIGURE 5.8
Signal flow graph for a general two-port.

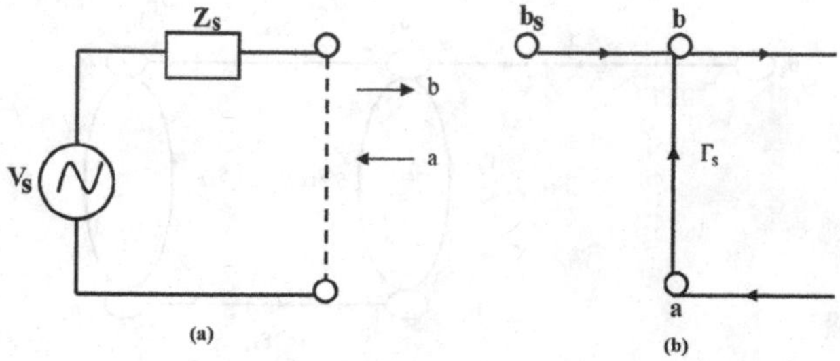

FIGURE 5.9
Signal generator (a) and its flow graph (b).

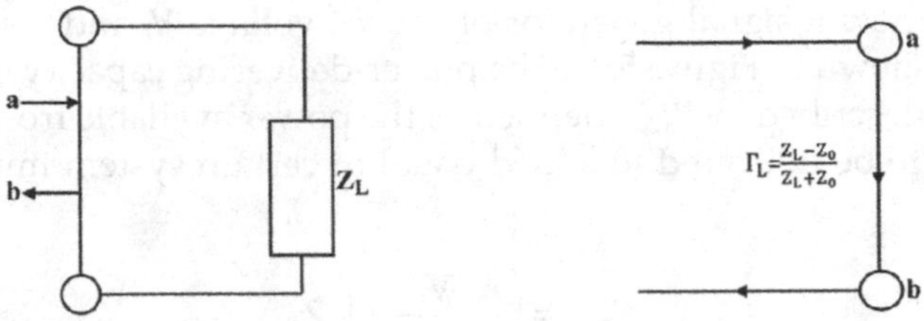

FIGURE 5.10
Load and its flow graph.

Next, we consider the application of a signal generator to a two-port network loaded by Z_L. The signal flow graph for such a system can be arrived at by combining the last three figures, as shown in Figure 5.11. It is noted that out of the five "power" variables shown in Figure 5.12 (b_s, a_1, b_1, a_2, and b_2), only b_s is independent. Signal flow graphs such as that shown in Figure 5.11 can be used to help evaluate the transfer functions and driving point immittance functions.

As an example, let us evaluate the power ration b_2/b_s of the circuit shown in Figure 5.11. This quantity may be interpreted as the ratio of the power incident to the load and the power available from the source. Figure 5.11 can be redrawn such that b_s is the input and b_2 is the output of the system, as shown in Figure 5.12. The signal flow diagram is reduced successively using the basic canonical form in the feedback system as indicated in Figure 5.12.

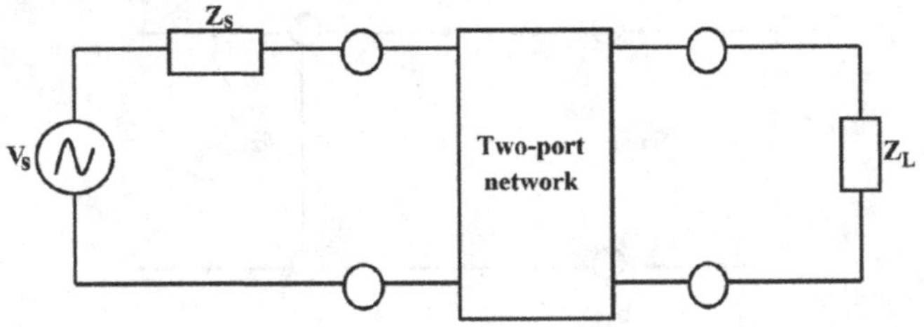

FIGURE 5.11
Loaded two-port network with source.

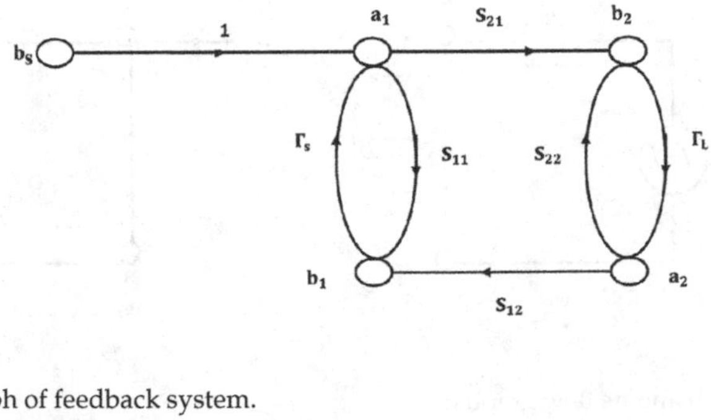

FIGURE 5.12
Signal flow graph of feedback system.

5.11 Derivation of G_T

$$\text{Transducer Power Gain} \quad G_T = \frac{P_L}{P_{AVS}} = \frac{\text{Power delivered to the load}}{\text{Power available from the source}} \tag{5.68}$$

$$\text{Operating Power Gain} \quad G_P = \frac{P_L}{P_{IN}} = \frac{\text{Power delivered to the load}}{\text{Power input to the network}} \tag{5.69}$$

$$\text{Available Power Gain} \quad G_A = \frac{P_{AVN}}{P_{AVS}} = \frac{\text{Power available from the network}}{\text{Power available from the source}} \tag{5.70}$$

See Figure 5.13.

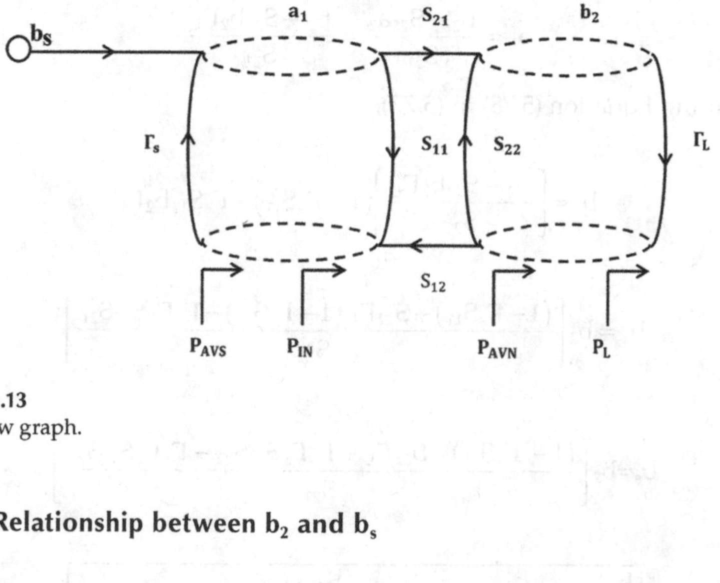

FIGURE 5.13
Single flow graph.

5.11.1 Relationship between b_2 and b_s

$$b_1 = S_{11}a_1 + S_{21}a_2 \tag{5.71}$$

$$b_2 = S_{21}a_1 + S_{22}a_2 \tag{5.72}$$

$$a_1 = b_s + \Gamma_s b_1 \tag{5.73}$$

$$a_2 = b_2\Gamma_L \tag{5.74}$$

From Equation (5.73),

$$b_s = a_1 - \Gamma_s b_1 \tag{5.75}$$

Substituting Equation (5.71) in (5.75),

$$b_s = a_1 - \Gamma_s S_{11} a_1 - \Gamma_s S_{12} a_2 \tag{5.76}$$

Substituting Equation (5.74) in (5.76),

$$b_s = a_1 - \Gamma_s S_{11} a_1 - \Gamma_s S_{12} b_2 \Gamma_L$$

$$b_s = a_1 (1 - \Gamma_s S_{11}) - \Gamma_s S_{12} b_2 \Gamma_L \tag{5.77}$$

From Equation (5.72),

$$a_1 = \frac{b_2 - S_{22} a_2}{S_{21}} = \frac{b_2 - S_{22} b_2 \Gamma_L}{S_{21}} \tag{5.78}$$

Substituting Equation (5.78) in (5.77),

$$b_s = \left[\frac{b_2 - S_{22} b_2 \Gamma_L}{S_{21}}\right](1 - \Gamma_s S_{11}) - \Gamma_s S_{12} b_2 \Gamma_L$$

$$b_s = b_2 \left[\frac{(1 - \Gamma_s S_{11}) - S_{22}\Gamma_L (1 - \Gamma_s S_{11}) - \Gamma_s \Gamma_L S_{12} S_{21}}{S_{21}}\right]$$

$$b_s = b_2 \left[\frac{(1 - \Gamma_s S_{11}) - S_{22}\Gamma_L + \Gamma_s \Gamma_L S_{11} S_{22} - \Gamma_s \Gamma_L S_{12} S_{21}}{S_{21}}\right]$$

$$\boxed{\frac{b_2}{b_s} = \frac{S_{21}}{(1 - \Gamma_s S_{11}) - S_{22}\Gamma_L + \Gamma_s \Gamma_L S_{11} S_{22} - \Gamma_s \Gamma_L S_{12} S_{21}}} \tag{5.79}$$

5.11.2 Evaluation of b_2/b_s by the Signal Flow Graph

See Figure 5.14.

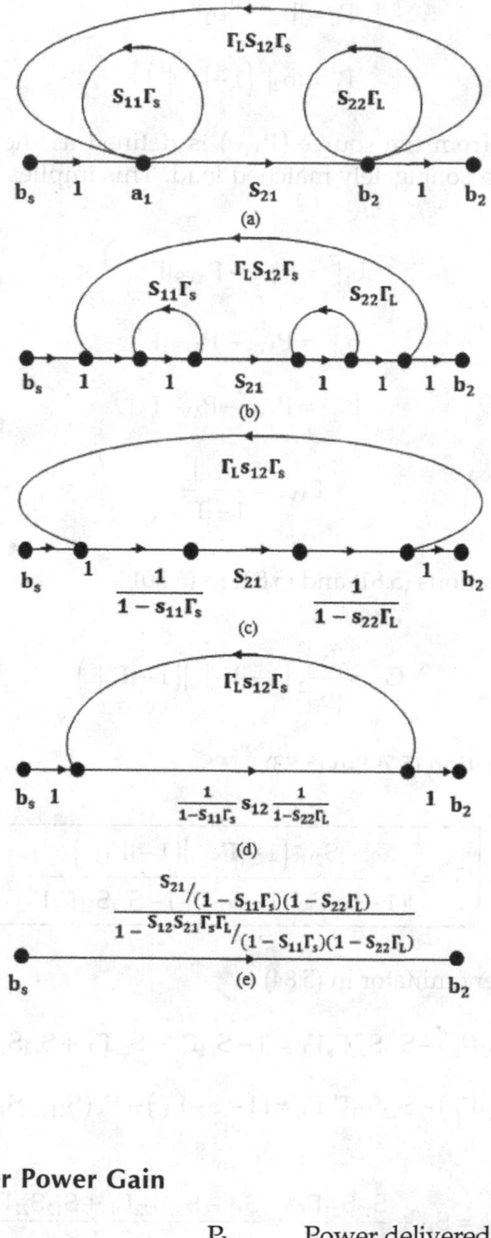

FIGURE 5.14
Evaluation of b_2/b_s.

5.11.3 Transducer Power Gain

$$\text{Transducer Power Gain} \quad G_T = \frac{P_L}{P_{AVS}} = \frac{\text{Power delivered to the load}}{\text{Power available from the source}} \tag{5.80}$$

Power delivered to the load is

$$P_L = |b_2|^2 - |a_2|^2$$

$$P_L = |b_2|^2 - |b_2|^2 |\Gamma_L|^2$$

$$P_L = |b_2|^2 \left(1 - |\Gamma_L|^2\right) \tag{5.81}$$

Power available from the source (P_{AVS}) is defined as the power delivered by the source to a conjugately matched load. This implies (from figure) that $\Gamma_{in} = \Gamma_s^*$

$$|b_s|^2 = P_{AVS} - P_{AVS}|\Gamma_{in}|^2$$

$$|b_s|^2 = P_{AVS} - P_{AVS}|\Gamma_s^*|^2$$

$$|b_s|^2 = P_{AVS} - P_{AVS}|\Gamma_s|^2$$

$$P_{AVS} = \frac{|b_s|^2}{1 - |\Gamma_s|^2} \tag{5.82}$$

Substituting Equations (5.81) and (5.82) in (5.80),

$$G_T = \frac{|b_2|^2}{|b_s|^2}\left(1 - |\Gamma_L|^2\right)\left(1 - |\Gamma_s|^2\right) \tag{5.83}$$

Substituting Equation (5.79) in (5.83),

$$\boxed{G_T = \frac{|S_{21}|^2\left(1 - |\Gamma_S|^2\right)\left(1 - |\Gamma_L|^2\right)}{\left|(1 - S_{11}\Gamma_S)(1 - S_{22}\Gamma_L) - S_{12}S_{21}\Gamma_S\Gamma_L\right|^2}} \tag{5.84}$$

Simplying the denominator in (5.84)

$$(1 - S_{11}\Gamma_s)(1 - S_{22}\Gamma_L) - S_{12}S_{21}\Gamma_s\Gamma_L = 1 - S_{11}\Gamma_s - S_{22}\Gamma_L + S_{12}S_{21}\Gamma_s\Gamma_L + S_{11}S_2\Gamma_s\Gamma_L$$

$$(1 - S_{11}\Gamma_s)(1 - S_{22}\Gamma_L) - S_{12}S_{21}\Gamma_s\Gamma_L = (1 - S_{22}\Gamma_L) - \Gamma_s(S_{11} - S_{11}S_{22}\Gamma_L + S_{12}S_{21}\Gamma_L) \tag{5.85}$$

We know that $\Gamma_{in} = S_{11} + \dfrac{S_{12}S_{21}\Gamma_L}{1 - S_{22}\Gamma_L} = \dfrac{S_{11} - S_{11}S_{22}\Gamma_L + S_{12}S_{21}\Gamma_L}{1 - S_{22}\Gamma_L}$

$$\therefore S_{11} - S_{11}S_{22}\Gamma_L + S_{12}S_{21}\Gamma_L = \Gamma_{in}(1 - S_{22}\Gamma_L) \tag{5.86}$$

Substituting Equation (5.86) in (5.85),

$$\therefore (1-S_{11}\Gamma_s)(1-S_{22}\Gamma_L)-S_{12}S_{21}\Gamma_s\Gamma_L=(1-S_{22}\Gamma_L)-\Gamma_s\Gamma_{in}(1-S_{22}\Gamma_L) \quad (5.87)$$

Substituting Equation (5.87) in (5.84),

$$G_T=\frac{|S_{21}|^2\left(1-|\Gamma_S|^2\right)\left(1-|\Gamma_L|^2\right)}{\left|(1-S_{22}\Gamma_L)-\Gamma_s\Gamma_{in}(1-S_{22}\Gamma_L)\right|^2}$$

$$G_T=\frac{|S_{21}|^2\left(1-|\Gamma_S|^2\right)\left(1-|\Gamma_L|^2\right)}{\left|(1-S_{22}\Gamma_L)(1-\Gamma_s\Gamma_{in})\right|^2}$$

or

$$\boxed{G_T=\frac{1-|\Gamma_S|^2}{|1-\Gamma_{in}\Gamma_s|^2}|S_{21}|^2\frac{1-|\Gamma_L|^2}{|1-S_{22}\Gamma_L|^2}} \quad (5.88)$$

Similarly, for the output reflection coefficient (Γ_{out}),

$$\boxed{G_T=\frac{1-|\Gamma_S|^2}{|1-S_{11}\Gamma_S|^2}|S_{21}|^2\frac{1-|\Gamma_L|^2}{|1-\Gamma_{OUT}\Gamma_L|^2}} \quad (5.89)$$

- When the power delivered to the load (P_L) is equal to the power available from the network (P_{AVN}), transducer power gain (G_T) becomes available power gain (G_A), that is, $G_T = G_A$.

$$\boxed{G_A=\frac{1-|\Gamma_S|^2}{|1-S_{11}\Gamma_S|^2}|S_{21}|^2\frac{1}{1-|\Gamma_{OUT}|^2}} \quad \left[\because P_L=P_{AVN} \quad \text{when} \quad \Gamma_L=\Gamma_{out}^*\right] \quad (5.90)$$

- When the power available from the source (P_{AVS}) is equal to the power input to the network (P_{IN}),the transducer power gain (G_T) becomes the operating power gain (G_P), that is, $G_T = G_P$.

$$\boxed{G_P=\frac{1}{1-|\Gamma_{IN}|^2}|S_{21}|^2\frac{1-|\Gamma_L|^2}{|1-S_{22}\Gamma_L|^2}} \quad \left[\because P_L=P_{AVS} \quad \text{when} \quad \Gamma_{In}=\Gamma_s^*=\Gamma_{out}^*\right] \quad (5.91)$$

$$G_T=\frac{1-|\Gamma_S|^2}{|1-\Gamma_{IN}\Gamma_S|^2}|S_{21}|^2\frac{1-|\Gamma_L|^2}{|1-S_{22}\Gamma_L|^2}$$

$$G_T = \frac{1-|\Gamma_S|^2}{|1-S_{11}\Gamma_S|^2}|S_{21}|^2\frac{1-|\Gamma_L|^2}{|1-\Gamma_{OUT}\Gamma_L|^2}$$

$$G_P = \frac{1}{1-|\Gamma_{IN}|^2}|S_{21}|^2\frac{1-|\Gamma_L|^2}{|1-S_{22}\Gamma_L|^2}$$

$$G_A = \frac{1-|\Gamma_S|^2}{|1-S_{11}\Gamma_S|^2}|S_{21}|^2\frac{1}{1-|\Gamma_{OUT}|^2}$$

5.11.4 Matched Transducer Power Gain (G_{TM})

$(\Gamma_S = \Gamma_L = 0)$ when both the input and output networks are perfectly matched.

$$G_{TM} = |S_{21}|^2 \tag{5.92}$$

5.11.5 Unilateral Transducer Power Gain (G_{TU})

G_{TU} is the forward power gain in a feedback amplifier, having its reverse power gain set to zero $\left(|S_{12}|^2 = 0\right)$ by adjusting the lossless reciprocal feedback network connected around the microwave amplifier.

$$G_{TU} = \frac{1-|\Gamma_S|^2}{|1-S_{11}\Gamma_S|^2}|S_{21}|^2\frac{1-|\Gamma_L|^2}{|1-S_{22}\Gamma_L|^2} \tag{5.93}$$

5.11.6 Maximum Unilateral Transducer Power Gain ($G_{TU_{max}}$)

The maximum unilateral transducer power gain is obtained when $\Gamma_s = S_{11}^*$ and $\Gamma_L = S_{22}^*$.

$$G_{TU_{max}} = \frac{|S_{21}|^2}{\left(1-|S_{11}|^2\right)\left(1-|S_{22}|^2\right)} \tag{5.94}$$

5.12 Reflection Coefficients in Terms of S Parameters

$$\text{Unilateral Transducer Power Gain } (G_{TU}) = \frac{1-|\Gamma_s|^2}{|1-S_{11}\Gamma_s|^2}|S_{21}|^2\frac{1-|\Gamma_L|^2}{|1-S_{22}\Gamma_L|^2} \tag{5.95}$$

$$\text{Source Gain}\,(G_s) = \frac{1-|\Gamma_S|^2}{|1-S_{11}\Gamma_S|^2} \tag{5.96}$$

$$\text{Load Gain}\,(G_T) = \frac{1-|\Gamma_L|^2}{|1-S_{22}\Gamma_L|^2} \tag{5.97}$$

These gains are maximized when

$$\Gamma_s = S_{11}^* \text{ and } \Gamma_L = S_{22}^* \tag{5.98}$$

$$\therefore G_{S_{max}} = \frac{1}{1-|S_{11}|^2} \tag{5.99}$$

$$G_{L_{max}} = \frac{1}{1-|S_{22}|^2} \tag{5.100}$$

5.13 Normalized Gain Factors g_s and g_L

$$g_s = \frac{G_s}{G_{S_{max}}} = \frac{1-|\Gamma_s|^2}{|1-S_{11}\Gamma_s|^2}\left(1-|S_{11}|^2\right) \tag{5.101}$$

$$g_L = \frac{G_L}{G_{L_{max}}} = \frac{1-|\Gamma_L|^2}{|1-S_{22}\Gamma_L|^2}\left(1-|S_{22}|^2\right) \tag{5.102}$$

Then, we have $0 \le g_s \le 1, 0 \le g_L \le 1$ for fixed values of g_s and g_L; the preceding two equations represent circles in the Γ_s or Γ_L plane.

From Equation (5.101),

$$g_s|1-S_{11}\Gamma_s|^2 = \left(1-|\Gamma_s|^2\right)\left(1-|S_{11}|^2\right)\left(g_s|S_{11}|^2+1-|S_{11}|^2\right)|\Gamma_s|^2 - g_s\left(S_{11}\Gamma_s + S_{11}^*\Gamma_s^*\right)$$

$$= 1-|S_{11}|^2 - g_s\Gamma_s\Gamma_s^* - \frac{g_s\left(S_{11}\Gamma_s + S_{11}^*\Gamma_s^*\right)}{1-(1-g_s)|S_{11}|^2}$$

$$= \frac{1-|S_{11}|^2 - g_s}{1-(1-g_s)|S_{11}|^2}$$

Adding $\left(g_s{}^2|S_{11}|^2\right)/\left(1-(1-g_s)|S_{11}|^2\right)^2$ to both sides to complete the square,

$$\left|\Gamma_s - \frac{g_s S_{11}^*}{1-(1-g_s)|S_{11}|^2}\right|^2 = \frac{\left(1-|S_{11}|^2 - g_s\right)\left[1-(1-g_s)|S_{11}|^2\right] + g_s^{\,2}|S_{11}|^2}{\left[1-(1-g_s)|S_{11}|^2\right]^2} \tag{5.103}$$

Simplifying the preceding equation,

$$\left|\Gamma_s - \frac{g_s S_{11}^*}{1-(1-g_s)|S_{11}|^2}\right|^2 = \frac{\sqrt{1-g_s}\left(1-|S_{11}|^2\right)}{1-(1-g_s)|S_{11}|^2} \tag{5.104}$$

$$\text{Center:}\quad C_s = \frac{g_s S_{11}^*}{1-(1-g_s)|S_{11}|^2} \tag{5.105}$$

$$\text{Radius:}\quad R_s = \frac{\sqrt{1-g_s}\left(1-|S_{11}|^2\right)}{1-(1-g_s)|S_{11}|^2} \tag{5.106}$$

Similarly, for load side,

$$\text{Center:}\quad C_L = \frac{g_L S_{22}^*}{1-(1-g_L)|S_{22}|^2} \tag{5.107}$$

$$\text{Radius:}\quad R_s = \frac{\sqrt{1-g_L}\left(1-|S_{22}|^2\right)}{1-(1-g_L)|S_{22}|^2} \tag{5.108}$$

5.14 Input Reflection Coefficient (Γ_{in})

We know that

$$b_1 = S_{11}a_1 + S_{21}a_2 \tag{5.109}$$

$$b_2 = S_{21}a_1 + S_{22}a_2 \tag{5.110}$$

$$a_1 = b_s + \Gamma_s b_1 \tag{5.111}$$

$$a_2 = b_2\Gamma_L \tag{5.112}$$

$$\Gamma_{in} = \frac{b_1}{a_1} \tag{5.113}$$

Substituting Equation (5.109) in (5.113),

$$\Gamma_{in} = \frac{b_1}{a_1} = \frac{S_{11}a_1 + S_{21}a_2}{a_1} = S_{11} + S_{12}\left(\frac{a_2}{a_1}\right)$$

$$\Gamma_{in} = S_{11} + S_{12}\left(\frac{a_2}{b_2} \times \frac{b_2}{a_1}\right)$$

$$\Gamma_{in} = S_{11} + S_{12}\Gamma_L\left(\frac{b_2}{a_1}\right) \quad \left(\because \Gamma_L = \frac{a_2}{b_2}\right) \tag{5.114}$$

Substituting Equation (5.112) in (5.110),

$$b_2 = S_{21}a_1 + S_{22}b_2\Gamma_L$$

$$b_2 - S_{22}b_2\Gamma_L = S_{21}a_1$$

$$b_2(1 - S_{22}\Gamma_L) = S_{21}a_1$$

$$\frac{b_2}{a_1} = \frac{S_{21}}{1 - S_{22}\Gamma_L} \tag{5.115}$$

Substituting Equation (5.115) in (5.114),

$$\boxed{\Gamma_{in} = S_{11} + \frac{S_{12}S_{21}\Gamma_L}{1 - S_{22}\Gamma_L}} \tag{5.116}$$

5.15 Output Reflection Coefficient (Γ_{out})

$$\Gamma_{out} = \frac{b_2}{a_2} = \frac{S_{21}a_1 + S_{22}a_2}{a_2} = S_{22} + S_{21}\left(\frac{a_1}{a_2}\right) \quad (\because \text{from Equation 5.108}) \tag{5.117}$$

$$\Gamma_{out} = S_{22} + S_{21}\left(\frac{a_1}{b_1} \times \frac{b_1}{a_2}\right)$$

$$\Gamma_{out} = S_{22} + S_{21}\Gamma_s\left(\frac{b_1}{a_2}\right) \quad \left(\because \Gamma_s = \frac{a_1}{b_1}\right) \tag{5.118}$$

Substituting Equation (5.111) in (5.109),

$$b_1 = S_{11}\Gamma_s b_1 + S_{12}a_2 \quad (\because \text{for } \Gamma_{out}\ b_s = 0)$$

$$b_1 - S_{11}\Gamma_s b_1 = S_{12}a_2$$

$$b_1(1 - S_{11}\Gamma_s) = S_{12}a_2$$

$$\frac{b_1}{a_2} = \frac{S_{12}}{(1 - S_{11}\Gamma_s)} \tag{5.119}$$

Substituting Equation (5.119) in (5.118) (Table 5.1),

TABLE 5.1

Microwave Oscillator Design is Similar to Microwave Amplifier Design

	Oscillator	Amplifier
Similarity	• Same DC biasing circuit • Same active device • Same set of S-parameters • Negative resistance characteristics	
Difference	It has a tuning mechanism for its oscillation	It has an AC signal as its input source
	Compressed Smith chart is used for the design	Normal Smith chart is used for the design
	Γ_{in} and $\Gamma_{out} > 1$	Γ_{in} and $\Gamma_{out} < 1$
	Generator tuning network (determines negative resistance oscillating frequency)	Input matching circuit (input circuit at the source port)
	Load matching network (provides matching function)	Output matching circuit (output circuit at the load port)

$$\boxed{\Gamma_{out} = S_{22} + \frac{S_{12}S_{21}\Gamma_s}{1 - S_{11}\Gamma_s}} \tag{5.120}$$

5.16 Summary

See Figure 5.15.

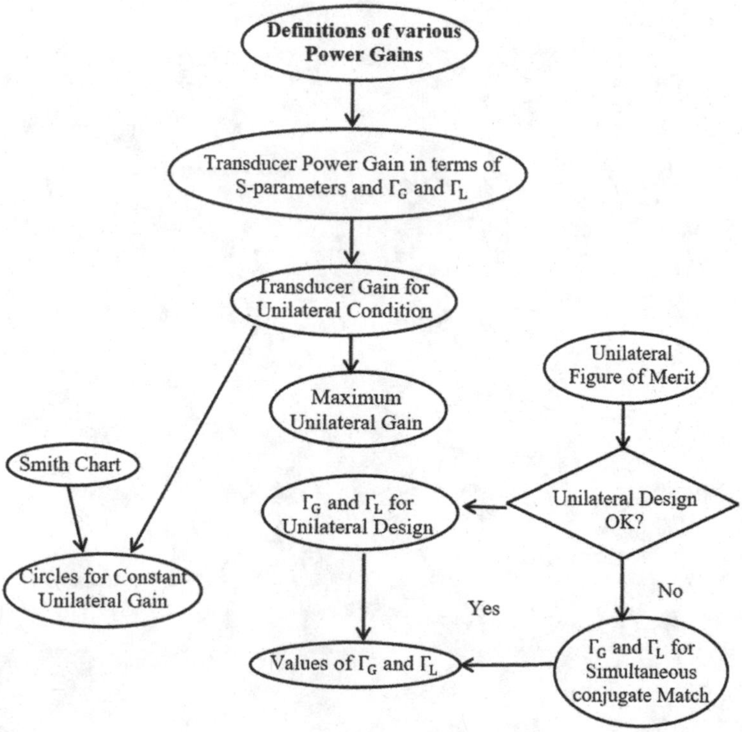

FIGURE 5.15
Flow of concepts needed for evaluation of the reflection coefficients Γ_G and Γ_L that the transistor needs to see at the input and output ports, respectively.

Bibliography

Samuel Y. Liao (1998), *Microwave Amplifiers and Circuits,* Prentice-Hall International Inc., Englewood Cliffs, NJ.

Appendix I: Ten Commandments for Microwaves (Thomas Laverghetta)

I. Thou shalt always remember that microwaves are very high frequencies and require special care.

II. Thou shalt always be aware of circuit mismatches and keep thy VSWRs very low.

III. Thou shalt be aware of what connector thy neighbor's components have on them so that adapters can be eliminated.

IV. Thou shalt always think of grounds in terms of planes and not as points. Thou shalt use points only on your pencil or compass.

V. Thou shalt not exceed the power rating of your equipment or components.

VI. Thou shalt check thy cables to be sure they are of the proper impedance, knowing fully well that RG-62A/U is not a 50 Ω cable.

VII. Thou shalt always remember that RF energy can burn, and whenever working with it, thou should repeat to thyself—"Thou shalt not touch."

VIII. Thou shalt lay out microwave components with great care, knowing that spacings and heights are critical.

IX. Thou shalt never be accused of turning in an improper data sheet when testing microwave components.

X. Thou shalt always maintain a sense of humor, knowing fully well that microwaves are not black magic but a science that few people can do without.

Numerical Techniques for Microwave and mm Wave Passive Structures

1. Finite Difference Method
2. Finite Element Method (FEM)
3. TLM Method (Transmission Line Matrix)
4. Integral Equation Method
5. Moment Methods and Galerkin's Method

6. Mode Matching Method
7. Transverse Resonance Technique
8. Method of Lines
9. Generalized Scattering Matrix Method
10. Spectral Domain Method
11. Equivalent Waveguide Method
12. Planar Circuit Model

Comparison of Numerical Methods

Method	Storage Requirement	CPU Time	Generality	Preprocessing
Finite difference	L	L	VG	Nil
FEM	L	ML	VG	S
Boundary element	M	M	VG	S
Transmission line matrix	ML	ML	VG	S
Integral equation	SM	SM	G	M
Mode matching	M	SM	G	M
Transverse resonance	SM	SM	Ma	M
Matching of lines	M	S	G	L
Spectral domain	S	S	Ma	L

L, large; M, moderate; S, small; VG, very good; G, good; Ma, marginal.

Typically Realizable Components in MICs

Passive Components

1. Low Pass, high pass
2. Band Pass, band stop with medium to large bandwidths
3. Directional couplers
4. Power splitters
5. Power combiners
6. Circulators, isolators
7. Attenuators
8. Resonators

9. Transformer circuits
10. Branches
11. Antennas

Semiconductor Circuits

1. Demodulators
2. Receiver and up-converter mixers
3. FET and BJT amplifiers
4. Transistor power amplifiers
5. Frequency multipliers
6. PIN-diode switches
7. PIN-diode phase shifters
8. Limiters
9. Modulators
10. Oscillators with transistors, Gunn and IMPATT diodes
11. Controllable attenuators

Subsystems

1. High-frequency receiver components
2. Transmitters (small signal and pulsed transmitter)
3. Phase-controlled antenna modules
4. Doppler radar

Inventions and Year

1. The telegraph (using Morse code) (1844)
2. Invention of the telephone (1876)
3. Radio communication by voice (1906)
4. First BCC TV broadcast (1936)
5. Birth of the Internet (1969)
6. Mobile telephones (1986)
7. The first transatlantic telegraph code (1866)
8. Invention of the wireless by Marconi (1887)
9. Invention of television (1923)

10. First electronic computer and transistor invented (1948)
11. First microprocessor (1971)
12. World Wide Web (1988)

Evolution of Ungrounded Antenna

1. Hertz Resonator—1888
2. Lodge's Resonant Antenna—1898
3. Brown's Slope Antenna—1902
4. Dipole Antenna—1900
5. Shielded Loop—1921
6. Yagi–Uda Antenna—1926
7. Chireix Mesny Array—1929
8. Rhombic Antenna—1931
9. Helical Antenna—1947
10. Turnstile Antenna—1936

Evaluation of Aperture Antenna

1. Hertz Parabolic Mirror—1888
2. Marconi's Parabolic Antenna—1933
3. Horn—1935
4. Parabolic Reflector—1935
5. Horn Reflector Antenna—1948

Some of the More Popular Commercial EM Software Packages

Ansoft HFSS—Finite Element Method (FEM) by Silvester

Full Wave—Finite Element Method

Ensemble—Method of Moments (MoM) by R. F. Harrington
Momentum—MoM
Em Suite—MoM
IE3D—MoM
Fidelity—Finite Difference Time Domain (FDTD) by S. Yee
Empire—FDTD
Concerto—FDTD
XFDTD—FDTD
EMA3D—FDTD
Microwave Studio—Perfect Boundary approximation
EM sight—Spectral Domain Moment Method
EM power—Method of Lines
MEFiSTo-2D—Transmission Line Method
Mafia—Finite integration Method

ESSOF

Libra—Nonlinear steady-state frequency domain simulation
Touchstone—Linear frequency domain circuit simulation
E-SYN—Automated matching Network Synthesis and Filter Design
LineCalc—Transmission line analysis and synthesis
Microwave SPICE—Nonlinear Time Domain simulation
ACADEMY—Graphical Design Environment for EEsof Simulators.
Omnisys—RF/Microwave system and subsystem analysis
Jomega—Linear/Nonlinear simulation of RF circuits
ANACAT—Computer-aided test
Xtract—Linear/Nonlinear characterization of microwave and RF GaAs FET and BJT Models.
FDTD (Finite Difference Time Domain)—Yee
FEM (Finite Element Method)—Silvester
Mode Matching Method—Wexler
Solution Techniques in Finite Element Method
Method of Moments (MoM)—Harrington
MoM for Implementation in Planar Simulator—Rautio and Harrington.
Transmission Line Matrix (TLM)—P.B. Johno

Three important reasons to simulate RF and microwave circuits and systems

- To understand the physics of a complex system of interacting elements.
- To test new concepts.
- To optimize designs.

As the frequency of RF circuits extends beyond a GHz to tens and hundreds of GHz, wavelengths become large with respect to the device and circuit dimensions, and the three-dimensional electromagnetic environment becomes more significant.

CAD Tools for Antenna Analysis

Some Commercially Available Microstrip Antenna CAD Tools

Software Name	Theoretical Model	Company
Ensemble (Designer)	Moment method	Ansoft
IE3D	Moment method	Zeland
Momentum	Moment method	HP
EM	Moment method	Sonnet
PiCasso	Moment method/genetic	EMAG
FEKO	Moment method	EMSS
PCAAD	Cavity model	Antenna Design Associates, Inc.
Micropatch	Segmentation	Microstrip Designs, Inc.
Microwave Studio (MAFIA)	FDTD	CST
Fidelity	FDTD	Zeland
HFSS	Finite element	Ansoft

Summary of Microwave Integrated Circuit

Why S-Parameters?

1. The equipment to measure total voltage and total current at the ports of the network is not readily available.
2. Short and open circuits are difficult to achieve over a broadband of frequencies because of lead inductance and capacitance.

3. Active devices such as transistors and negative diodes are very often not short- or open-circuit stable.
4. Voltages and current are not well-defined for a microwave circuit.
5. It is possible to define voltage as being proportional to transverse electric (TE) fields and current as being proportional to transverse magnetic (TM) fields.
6. A more satisfactory approach is to use incident and reflected waves as variables.
7. The values of h, y, or z parameters, ordinarily used in circuit design at lower frequencies, cannot be measured accurately above 100 MHz, because establishment of the required short and open circuit conditions is difficult.
8. They are derived from power ratios and consequently provide a convenient method for measuring circuit losses.
9. They are easy to measure because they are based on reflection characteristics rather than short or open circuit parameters.
10. They can make the designer's job easier.
    Scattering parameters simplify the design.

Microwave Measurements

1. $f_c(\text{Critical frequency}) = \dfrac{1}{2\sqrt{\mu\varepsilon}}\sqrt{\left(\dfrac{m}{a}\right)^2 + \left(\dfrac{n}{b}\right)^2}$

$$f_c\,(\text{air}) = \frac{c}{2}\sqrt{\left(\frac{m}{a}\right)^2 + \left(\frac{n}{b}\right)^2}$$

2. $\lambda_c(\text{cutoff wavelength}) = \dfrac{2}{\sqrt{\left(\dfrac{m}{a}\right)^2 + \left(\dfrac{n}{b}\right)^2}}$

3. $\beta\,(\text{during propagation}) = \sqrt{\omega^2\mu\varepsilon - \left(\dfrac{m}{a}\right)^2 - \left(\dfrac{n}{b}\right)^2}$

$$\beta\,(\text{in air during propagation}) = \frac{2\pi f}{c}\sqrt{1-\left(\frac{f_c}{f}\right)^2}$$

$$\text{In free space} \quad \lambda = \frac{c}{f}$$

$$\beta = \frac{2\pi}{\lambda}\sqrt{1-\left(\frac{f_c}{f}\right)^2}$$

4. $\lambda_g = \dfrac{\lambda_0}{\sqrt{1-\left(\dfrac{f_c}{f}\right)^2}}$ (or) $\boxed{\dfrac{1}{{\lambda_0}^2} = \dfrac{1}{{\lambda_g}^2} + \dfrac{1}{{\lambda_c}^2}}$

5. $v_g = \dfrac{1}{\sqrt{\mu\varepsilon}\sqrt{1-\left(\dfrac{f_c}{f}\right)^2}}$

6. $v_g \text{ (in air)} = \dfrac{c}{\sqrt{1-\left(\dfrac{f_c}{f}\right)^2}}$

$$\text{In air } v_g = \frac{c^2}{v_p} = \frac{1}{\dfrac{\partial\beta}{\partial\omega}} = \frac{\lambda}{\sqrt{1-\left(\dfrac{f_c}{f}\right)^2}}$$

As one approaches the critical frequency from higher frequencies, the phase velocity $\left(v_p\right)$ and the wavelength in the guide become larger.

In a waveguide with an air dielectric,

$$\boxed{v_p v_g = c^2}$$

$$\text{Power } P = E_{max} I_{min}$$

$$I_{min} = \frac{E_{min}}{Z_0}$$

$$\therefore P = \frac{E_{max} E_{min}}{Z_0}$$

If there is no standing wave, as in a perfectly matched line, then

$$E_{max} = E_{min} = \left|E^+\right|$$

$$P = \frac{|E|}{Z_0}$$

The wavelengths at microwave frequencies are usually comparable with the physical size of the circuit components. Hence the concept of distributed circuits. Sir Oliver Lodge in 1886 used the word impedance.

For TEM mode propagation,

$$Z = V/I$$

$$Z = W/II^*$$

$$Z = VV^*/W$$

For waveguide mode propagation, wave impedance is

$$Z = \frac{E_t}{H_t}$$

E_t, H_t – Electric field intensity and magnetic field intensity transverse to the direction of propagation.

7. $Z_{TE} = \dfrac{\eta}{\sqrt{1-\left(\dfrac{\lambda}{\lambda_c}\right)^2}} = \dfrac{\eta}{\sqrt{1-\left(\dfrac{f_c}{\lambda}\right)^2}}$

$$Z_{TM} = \eta\sqrt{1-\left(\frac{\lambda}{\lambda_c}\right)^2} = \eta\sqrt{1-\left(\frac{f_c}{\lambda}\right)^2}$$

$$Z_W{}^{TM} = \frac{\beta}{\omega\varepsilon} = \sqrt{\frac{\mu}{\varepsilon}} \quad \text{for TM mode}$$

$$Z_W{}^{TE} = \frac{\omega\mu}{\beta} = \sqrt{\frac{\mu}{\varepsilon}} \quad \text{for TE mode}$$

where η – Intrinsic impedance of the transmission line medium (377 Ω for free space)

8. $\tan\delta = \dfrac{\text{Energy dissipated}}{\text{Energy stored}}$

9. Dissipated Factor $= \dfrac{\text{Loss (Resistance)}}{\text{Conductivity}}$

Antenna Characteristics and Parameters

1. Input Voltage Standing Wave Ratio and input impedance

$$\text{VSWR} = \frac{1+\Gamma}{1-\Gamma}$$

$$Z_{in} = Z_0\left(\frac{1+\Gamma}{1-\Gamma}\right)$$

$$\text{Return loss in dB} = -20\log_{10}|\Gamma|$$

$$\text{Noise Figure in dB} = 10\log\frac{S_i/N_i}{S_o/N_o}$$

2. Band width
3. Power radiation patterns
4. Half-power beam width and side lobe level
5. Directivity, gain, and efficiency

$$D(\theta,\Phi) = S(\theta,\Phi)/\left(P_t/4\pi R^2\right)$$

$$D_{max} = \text{Maximum of } S(\theta,\Phi)/\left(P_t/4\pi R^2\right)$$

where $S(\theta,\Phi) = \frac{1}{2}\text{Re}\left[\vec{E}\times\vec{H}\right]$

Gain $(G) = \eta D_{max}$

$$\eta = \text{Efficiency} = \frac{P_{rad}}{P_{in}} = \frac{P_{rad}}{P_{rad} + P_{loss}}$$

6. Polarization and cross-polarization level
7. Beam efficiency
8. Back radiation
9. Effective area (A_e)

$$G = \frac{4\pi}{{\lambda_0}^2}A_e$$

10. Estimation of high-gain antenna.

Effects of Discontinuities

1. Frequency shift in narrowband circuits.
2. Degradation in input and output VSWR.
3. Higher ripple in gain flatness of broadband Integrated Circuits.
4. Interfacing problem in multifunction circuits.
5. Lower circuit yield due to the degradation in circuit performance.
6. Surface wave and radiation couplings that may cause oscillations in high-gain amplifiers.

Microwave Amplifier Formulas

1. Load reflection coefficient $(\Gamma_L) = \dfrac{Z_L - Z_0}{Z_L + Z_0}$

2. Source reflection coefficient $(\Gamma_s) = \dfrac{Z_s - Z_0}{Z_s + Z_0}$

3. Input reflection coefficient $(\Gamma_{in}) = S_{11} + \dfrac{S_{12}S_{21}\Gamma_L}{1 - S_{22}\Gamma_L} = \dfrac{S_{11} - \Delta\Gamma_L}{1 - S_{22}\Gamma_L}$

4. Output reflection coefficient $(\Gamma_{out}) = S_{22} + \dfrac{S_{12}S_{21}\Gamma_s}{1 - S_{11}\Gamma_s} = \dfrac{S_{22} - \Delta\Gamma_s}{1 - S_{11}\Gamma_s}$

5. Available power gain $G_A = \dfrac{P_{AVN}}{P_{AVS}} = \dfrac{1 - |\Gamma_s|^2}{|1 - S_{11}\Gamma_s|^2}|S_{21}|^2 \dfrac{1}{1 - |\Gamma_{out}|^2}$

6. Transducer power gain $G_T = \dfrac{P_L}{P_{AVS}} = \dfrac{1 - |\Gamma_E|^2}{|1 - S_{11}\Gamma_S|^2}|S_{21}|^2 \dfrac{1 - |\Gamma_L|^2}{|1 - \Gamma_{out}\Gamma_L|^2}$

7. Unilateral Transducer Power Gain $(G_{TU}) = \dfrac{1 - |\Gamma_s|^2}{|1 - S_{11}\Gamma_s|^2}|S_{21}|^2 \dfrac{1 - |\Gamma_L|^2}{|1 - S_{22}\Gamma_L|^2}$

Smith Chart

- Most widely used Radio Frequency graphical design tool to display the impedance behavior of a transmission line as a function of either line length or frequency.

- $\left[\Gamma_r - \dfrac{r}{r+1}\right]^2 + {\Gamma_r}^2 = \left(\dfrac{1}{r+1}\right)^2$

$$[\Gamma_r - 1]^2 + \left[\Gamma_i - \frac{1}{x}\right]^2 = \left(\frac{1}{x}\right)^2$$

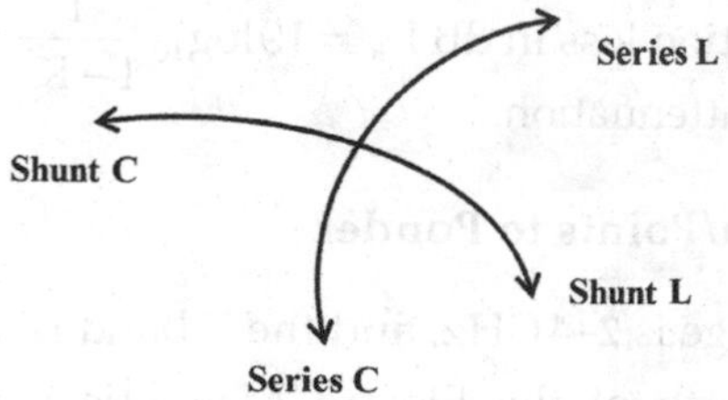

- $z = r + jx$

- $Q_n = \frac{|x|}{r} = \frac{2|\Gamma_i|}{1-\Gamma_r^2-\Gamma_i^2} = \frac{1-\Gamma_r^2-\Gamma_i^2}{(1-\Gamma_r^2)+\Gamma_i^2}$

$$+j\frac{2\Gamma_i}{(1-\Gamma_r^2)+\Gamma_i^2}\Gamma_i^2+\left(\Gamma_r \pm \frac{1}{Q_n}\right)^2 = 1+\frac{1}{Q_n^2}$$

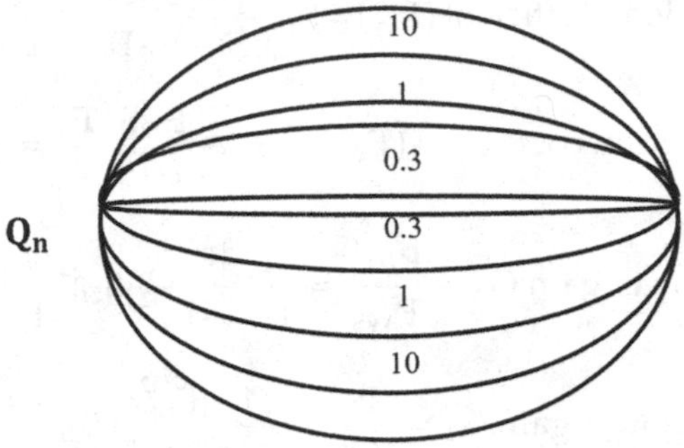

- LynnVille's chart (1956)—Modified form of the Smith chart is the formation of **Modern Solid-State Microwave Amplifier Design.**
- There are normally eight radial scales in addition to resistance and reactance circles on a standard Smith chart.
- Scale A & B—Both reflection coefficients.
  A—Distance from the center represents the absolute value of the voltage.
  B—Distance from the center represents the ratio of the reflective power to the incident power.
- G, H—Standing wave ratio. Distance from the center on scale H represents the VSWR. Distance from the center on scale G represents the VSWR in dB.
- Scale E—Return loss in dB $L_r = 20\log_{10}\frac{1}{K}$.
- Scale F—Reflection loss in dB $L_R = 10\log_{10}\frac{1}{1-K^2}$.
- Scale D—Line attenuation.

Useful Information/Points to Ponder

1. The S-band range is **2–4** GHz, and the X-band range is **8–12** GHz.
2. If the wavelength of the Electro Magnetic wave in a dielectric medium ($\varepsilon_r = 2$) is 2.12 mm, its corresponding free space wavelength is **2.998** mm.

3. The intrinsic impedance in a dielectric medium with a dielectric constant of $\varepsilon_r = 3$ is **217.5 Ω**.
4. 1 Kw corresponds to **60** dBm.
   100 W corresponds to **50 dBm.**
5. Would the VSWR be the same for load resistors of 100 and 25 Ω in a transmission line with a characteristic impedance of 50 Ω? **Yes**
6. Difference between 16 and 24 dBm is **–8 dBm.**
7. Input impedance at one-half wavelength from an open is **0 Ω (short).** Impedance at a point one-eight wavelength from a short resistive is **inductive** reactance.
8. The propagation in microstrip is the **quasi-Transverse Electro Magnetic (TEM)** mode.
9. The propagation in strip line is the **TEM** mode.
10. The return loss from a mismatch load is 2.95 dB. The reflection coefficient is **0.7120.**
11. The dominant mode in a rectangular waveguide is $\mathbf{TE_{10}}$.
12. The dominant mode in a circular waveguide is $\mathbf{TE_{11}}$.
13. If a rectangular wave is excited as a $\mathbf{TE_{10}}$ wave and is connected to a circular wave guide with proper interfacing, it will excite $\mathbf{TE_{11}}$ mode in the circular waveguide.
14. The diameter needed for a circular waveguide to operate in the $\mathbf{TE_{11}}$ mode at a frequency of 12 GHz is **0.0146 cm.** (The average Bessel root for $\mathbf{TE_{11}}$ mode is 1.84.)
15. Attenuation of the rotary vane attenuator if it is turned to an angle of 34° is **3.25 dB.**
16. The gain of an isotropic antenna is **0 dBi.**
17. The direction of the electric field of an electromagnetic wave radiated from an antenna is called **Polarization.**
18. The radiation from an antenna at other angles than the desired direction is known as **side lobe radiation.**
19. A major disadvantage of the microwave is that the **line of sight propagation attenuation increases after 60 km.**
20. Any device utilizing the temperature coefficient of resistivity of some λ resistance element is known as a **bolometer.**
21. Both **baretter** and **thermistor** are used for direction, and power level measurements fall in this category.
22. An E-H tee used for impedance transformation, having two arms terminated in adjustable plungers is known as **E-H tuner.**
23. The component used for rotating the plane of polarization is termed as **Ferrite device phase shifter (Fabry perot).**

24. For measuring high VSWR (3 dB method), what relation would you use for VSWR? Briefly mention what formula is obtained. $\lambda_g/\pi(d_1 - d_2)$
25. Ridged waveguide **lowers** the cutoff wavelength and **higher** frequency bandwidth.
26. The cutoff wavelength for circular waveguide depends on **transmission mode** and the **roots of Bessel equation.**
27. A helical antenna is **circularly** polarized.
28. The longest operating wavelength is defined as the **cutoff wavelength.**
29. The boundary conditions between the two dielectrics in cases of with and without charge on the interface and if there is charge on the interface are $E_{tan|die11} = E_{tan|die12} H_{orm|die11} = H_{norm|die12}$ and $E_{norm\,|\,die11} = E_{norm\,|\,die12}$, $H_{tan\,|\,die11} = H_{tan\,|\,die12}$, respectively.
30. The lowest resonant frequency of a cubic rectangular cavity with a side 5 cm (Express this frequency in terms of its corresponding free space wavelength) is **f = 4.239** GHz **and** λ **= 7 cm.**
31. The cutoff frequency of TEM mode is **0 GHz.**
32. **TEM** mode propagates through a cable.
33. TE_{01} cannot be considered as the dominant mode in a rectangular waveguide because the **cutoff frequency is greater.**
34. BNC stands for **Bayonet Neill–Concelman connector.**
35. ATC stands for **amphenol tube connector.**
36. Characteristic impedance is normally 50 Ω at high frequency because it is the **geometric mean (and also arithmetic mean) of 77 Ω (for optimum attenuation) and 30 Ω for optimum power.**
37. The range of microwave frequency is **3–30 GHz.**
38. Above **30** GHz is termed mm wave.
39. The band of operation normally in the educational intuitional microwave lab is **X-band.**
40. Guide wavelength will become equal to free space wavelength if $\varepsilon_r = 1$.
41. If a rectangular waveguide is filled with dielectric of ε_r, its cutoff wavelength is $\lambda_c/\sqrt{\varepsilon_r}$. λ_c **= cut off waveguide filled with air**
42. Of the two waveguides, which one will pass higher frequency?
43. Wave propagation through waveguides by means of **total internal reflection.**

✓(a) 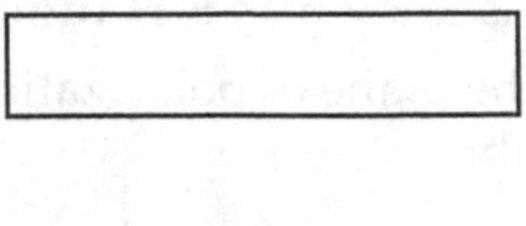

(b) 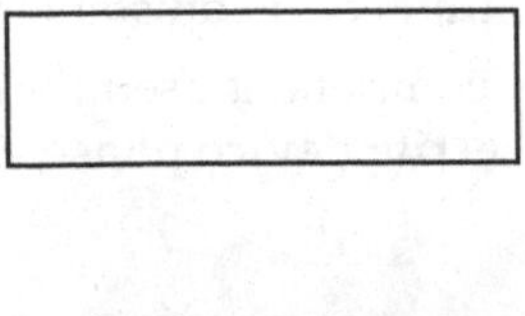

44. What is dielectric waveguide? What is its specialty?
    - Waveguide filled with dielectric material.
    - Lower losses
45. Draw TE_{20} mode, TM_{11} mode, and TE_{11} mode.
46. How to excite TE_{20} mode.
47. The mode excited in your microwave test bench is **TE_{10} mode.**
48. How the negative resistance characteristics is obtained in
    (a) Gunn diode, (b) **Tunnel diode**, (c) Impatt diode, (d) TRAPATT diode.
49. Give important differences between the traveling wave tube (TWT) and reflex klystron.

S. No	TWT	Reflex Klystron
1.	Nonresonant type	Resonant type
2.	High Q-value	Low Q-value
3.	External interaction of RF and electron beam	Local interaction of RF and electron beam

50. Klystron is named so because **klyzein in Greek means breaking of waves on the sea shore.**
51. What is the disadvantage of (a) Reflex klystron (b) TWT.
    (a) Bandwidth lesser and noise figure more
    (b) Bandwidth more and noise figure less.
52. Magnetron is called crossed field device because **electrical, magnetic waves cross each other in the cavities.**
53. Varactor can be regarded as low-noise device because **variable resistance absorbs noise power.**
54. The minimum distance to be kept in between two horn antennas while measuring gain is $2d^2/\lambda$ because of **interference.**
55. The free space impedance is **$120\pi\ \Omega$** and its value is obtained from $\sqrt{\mu_0/\varepsilon_0}$ **in free space propagation equation.**
56. To calculate guide wavelength, which distance will you take—the **difference between two successive minima** or that between two successive maxima.
57. Why is the successive minima difference taken as $\lambda/2$ and not λ?
58. What is PSWR? **Power Standing Wave Ratio** = $\left(\frac{E_{max}}{E_{min}}\right)^2$.
59. Give the expression to find out the input impedance of transmission line of characteristic impedance Z_0, terminating impedance Z_R, and length l is Z_{in}.

$$Z_{in} = Z_0\left[\frac{Z_R + Z_0 \tanh \gamma l}{Z_0 + Z_R \tanh \gamma l}\right] \quad \text{where } \gamma = \alpha + j\beta \text{ propagation constant}$$

60. If the terminating impedance is equal to its characteristic impedance, the input impedance is $Z_{in} = Z_0$ **(matched)**
61. Given an electrical equivalent of your frequency meter.
    LC tank resonant circuit
62. Write all Maxwell's equations.

$$\nabla \times E = -\frac{\partial B}{\partial t} \quad \nabla \times H = J + \frac{\partial D}{\partial t}$$

$$\nabla E = \frac{\rho}{t} \quad \nabla \times H = 0$$

63. Write your boundary conditions.

    1. $E_{tan|air} = E_{tan|dielectric}$
    2. $H_{norm|air} = H_{norm|dielectric}$
    3. $E_{norm\,|\,air} - E_{tan|dielectric} = \rho$
    4. $H_{tan|air} - H_{tan|dielectric} = J$

64. The maximum distance that microwave can travel without repeater is **60 km.**
65. If microwaves pass through ionosphere, what will happen?
    - It will get reflected.
66. How can different frequencies be adjusted in a reflex klystron?
    - Electronic Tuning and Mechanical Tuning
67. What is applegate diagram?
    - **In reflex klystron, the diagram in terms of** G_e **and** $B_{e.}$
68. What kind of oscillator is used in?
    (a) Microwave oven—**magnetron**
    (b) Microwave repeater stations—**Traveling Wave Tube (TWT)**
    (c) General purpose microwave lab—**gun diode/reflex klystron**
69. What is the other name for GaAS diode? Why is it named so?
    TED. Because negative characteristics are obtained by means of various domain transverse.
70. When was the first magnetron found? Can it be replaced in future?
    During Second World War (1939–1945). No, it cannot be replaced.

71. Can you give one application of biological effect of microwave?
    Treatment of cancer, burning away of cancerous tissues using microwaves.
72. What is microwave diathermy?
    Biological heating through microwave.
73. Is it possible for you to construct a Smith chart? If so, what is the basic equation?

$$K = \frac{Z_R - Z_0}{Z_R + Z_0}$$

74. What is the phase shift property of an S-parameter? What is its advantage?

$$[S] = [\phi][S][\phi]$$

where $[\phi] = \begin{bmatrix} e^{-j\beta l_1} & 0 \\ 0 & e^{-j\beta l_2} \end{bmatrix}$ wherever the length has to be modified.

75. Why do you provide isolation after gunn source?
    - For frequency stability.
    - This is to prevent the reflected waves of the system from going (hitting) back to the source.
76. Why do you provide attenuator in your test setup?
    - For the safety of measuring instruments.
    - The meters used are very sensitive. Slight change in power level can cause permanent damage. Hence, an attenuator is used.
77. What is the disadvantage of double-stub matching? How is it overcome?
    - All admittances cannot be matched, there is some forbidden region.
    - It can be overcome by adding one more stub (triple stub).
78. Which will have better propagation?
    (a) Gold-coated waveguide, (b) Silver-coated waveguide
79. Can you give an application of a circular waveguide?
    1. Rotary vane attenuators
    2. Cavity resonators.
80. Can you give an example of a nonreciprocal deceive?
    1. Isolator
    2. Circulator
    3. Gyrator using ferrite material.

81. Is it possible to have a perfect lossless three-port reciprocal circulator? **No**
82. What is the difference between a 3 dB hybrid and a magic tee?
    3-dB hybrid – Phase change
    Magic tee – No Phase change
83. What is immittance (inverted chart) Smith chart?
    It enables us to work with both impedance and admittance. It is actually impedance, admittance charts.
84. If a microwave system is good, then its VSWR is **1**. Its reflection coefficient is **0**.
85. Is it possible to have reflection coefficient more than 1? **Yes.** If so what Smith chart do you use for that? **Compressed Smith chart.**
86. Why is a PIN diode used in your gunn oscillator setup?
    - For modulation so that the measurement can be made possible.
87. If matched termination is connected at the end of a transmission line, what will be its VSWR? **Almost one (ideally 1).**
88. While calculating wavelength at a microwave frequency, why is 3×10^{10} cm taken into consideration ? **It is the velocity of light.**
89. The relationship between phase velocity, group velocity, and free space velocity is $c^2 = \upsilon_p \upsilon_g$.
90. Is it possible to find gain if two antennas are not identical? How?
    Yes. By using the equivalent antenna model in Friss transmission formula.
91. Why is square wave used as modulation in a reflex klystron setup?
    - To avoid continuous wave and also to have two levels (0 and 1) only.
92. What is elliptic polarization? $E_x \neq E_y$.
    When the E, H field components loci is the equation of an ellipse, the type of polarization is called elliptical polarization.
93. What is the value of directivity of an ideal directional coupler? What value do you normally get in the lab?
    Ideally ∞ Usual value 40 dB
94. What method is used for finding high VSWR?
    Double minimum method. (High VSWR > 10.)
95. Which method is preferred for finding high VSWR?
    Double minimum method.
96. Define anisotropy.
    Anisotropy is the nonuniform propagation of an electromagnetic wave due to varying values.
97. Give one example of anisotropy material. **Sapphire.**

98. What is poynting theorem?
    Total incident Power = Total dissipated power + stored electric energy + stored magnetic energy + power transmitted
99. Can TEM mode exist in a hollow waveguide. Why?
    No. This is because the absence of axial component in the hollow wave guide leads to rapid attenuation by the walls.
100. Draw an electrical equivalent circuit of a transmission line.

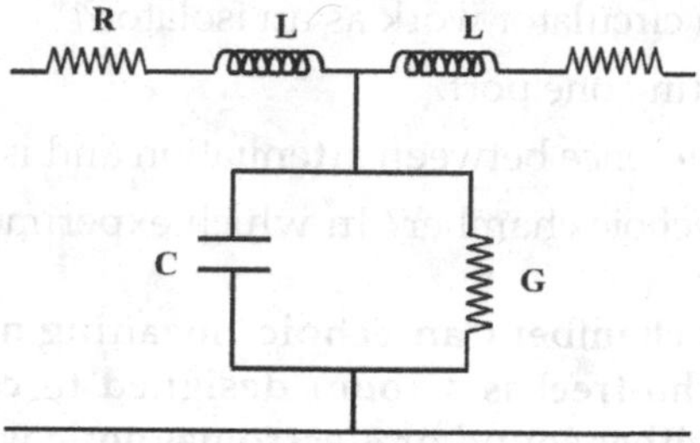

101. What voltage do you apply to gunn diode? What is the corresponding threshold voltage?
    Voltage applied (8–15 V), Threshold voltage 9 V
102. From directivity and coupling factor of a directional coupler, express its scattering matrix.

$$C = \sqrt{S_{13}}$$

$$D = \frac{S_{14}}{S_{13}}$$

103. Draw E-field in E-plane tee.

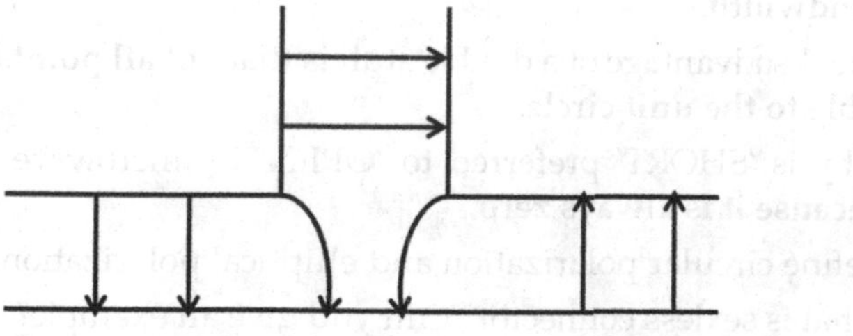

104. Why is the magic tee called so?
    By field
105. Express attenuation, isolation, reflection coefficient of a simple wave guide in terms of its S-parameters.

106. Express the following in dBm (a) 1 mW **0 dBm**
107. Express the following in dBw (a) 1 W **30 dBm**
108. Why is the slot provided at the center of a slotted section?
    - TE_{10} is excited. Hence, maximum field at the center.
109. What are the initial adjustments necessary before conducting an antenna experiment?
    - The distance between the two antennas is $2d^2/\lambda$.
110. How to make a circulator work as an isolator?
    - By terminating one port.
111. What is the difference between attenuation and isolation?
112. What is an anechoic chamber? In which experiment is it desired to be used?

    An anechoic chamber ("an-echoic" meaning nonreflective, non-echoing, or echo-free) is a room designed to completely absorb reflections of either sound or electromagnetic waves. It is used in antenna radiation pattern measurements.
113. What is Sabin's constant? Where is to be used?
114. A signal propagated in a waveguide has a full wave of electrical intensity change between the two further walls and no component of the electric field in the direction of propagation. The mode is $\mathbf{TE_{10}}$.
115. **PIN** diode is suitable for switching.
116. Impedance inversion may be obtained using the **$\lambda/4$ line.**
117. The one-eighth lambda transmission line transformer is useful to match the **load and source** if $|Z_L| = |Z_S|$
118. A half-wavelength transformer is useful in the transmission line because it will **repeat the property**.
119. The main disadvantage of a two-hole directional coupler is **lesser bandwidth.**
120. The disadvantage of a double stub is that **not all points are transferrable to the unit circle**.
121. Why is 'SHORT' preferred to 'OPEN' in microwave experiments? **Because it is always zero.**
122. Define circular polarization and elliptical polarization.
123. What is sexless connector? Can you give an example?

    It has no male, female joints. For example, Waveguide connector.
124. How is the rectangular waveguide designated (Similar to AC 128)? **WR 90. 90 is the wider dimension in mills.**
125. What is the use of the slide screw tuner? **To adjust the capacitance.**

126. Draw a rough figure of X-band short.

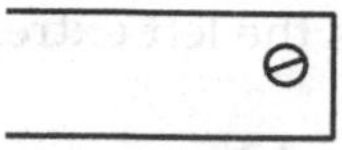

127. What is the impedance of termination? $\mathbf{Z_0}$
128. What is VSWR if the termination is (a) short (b) open.
   (a) Short ∞, (b) Open ∞
129. What is the difference between sector and horn antenna?
   Sector—one of the walls is flared.
   Horn—both walls are flared
130. What is path loss?
   The distance that microwaves can travel for a given atmospheric condition.
131. MESFET stands for **Microwave Semiconductor FET**.
132. The profile of a VSWR circle for a fixed load and different frequencies is a spiral.
133. The lowest frequency of propagation in a waveguide depends on **wider dimension** and that of highest frequency of propagation depends on **higher order modes**.
134. The relation between average power, peak power, and duty cycle is **Peak power = Average power/duty cycle.**
135. Bolo meters are two types, namely, **baretters** and **thermistor**, which have **positive** temperature coefficient and **negative** temperature.
136. The common types of microwave detectors are **crystal** and **bolometer**.
137. High-power measurement measures **about 10** W and the normal method is the **calorimetric method.**
138. Low-power measurement measures **lower than 1** W and the normal method is the **bolometric method.**
139. **Crystal** detectors generate a DC voltage when they absorb microwave power.
140. At microwave frequency it is difficult to measure **absolute voltage** and **current**. It is simpler to measure **standing wave ratio** directly.
141. Neper is defined as $\log_e \frac{P_1}{P_2}$.
142. In 3 dB hybrid (MIC), should the corners be cut or not? Why?
   Yes. To prevent reflections.
143. What is the use of high dielectric constant?
   To reduce the propagation wavelength.

144. Why is a fan provided near reflex klystron?
To act as a heat sink and cooling device.

145. In a VSWR meter, what is the left extreme marking and what is the right extreme marking?
Left end ∞ and right end 1

146. What are the types of attenuator? **Flap, variable, rotatory vane.**

147. Draw the gunn diode characteristics?

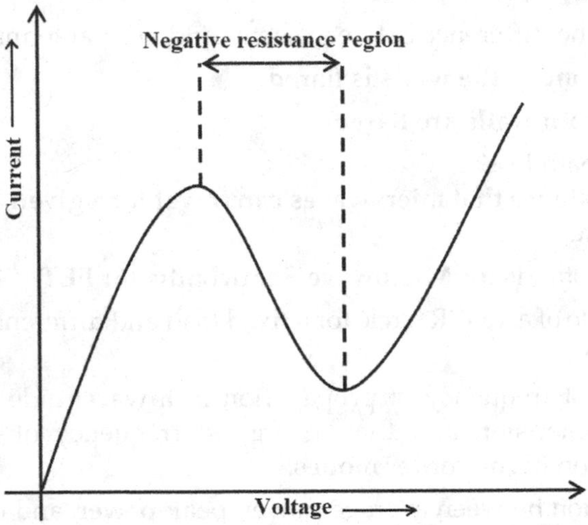

148. Does the operating frequency depend on the medium through which waves travel? **Yes**

149. Waveguide is a **high-pass** filter but all frequencies above cutoff frequency **will not be passed (unlike an ideal high pass filter).**

150. Draw the standing wave pattern for a matched load and that for a short.

151. Ideally, antenna measurements are to be conducted **in an anechoic chamber.**

152. The difference between sector antenna and horn antenna is the **smooth transition and abrupt transition.**

153. Define circular polarization and elliptical polarization.

154. Why is a PIN modulator provided?

155. What is the advantage of giving square modulation?

156. Define gain of a horn antenna.

157. If two antennas (T_x and R_x) are of different types, how will you do the antenna measurement?
Three antennas method.

Appendix II

Recommended Reading

Adams, T. M. and Layton, R. A., 2009. *Introductory MEMS: Fabrication and Applications*. Springer: New York.

Alan Davis, W., and Agarwal, K. K., 2003. *Radio Frequency Circuit Design*. John Wiley & Sons: New York.

Ananthasuresh, G. K., Vinoy, K. J., S. Gopalakrishnan, Bhat, K. N., and Aatre, V. K., 2012. *Micro and Smart Systems: Technology and Modeling*. John Wiley & Sons: New York.

Anritsu, 2008. *The Essentials of Vector Network Analysis: From α to Z_0*. Anritsu Company: Atsugi.

Bahl, I. J., 2003. *Lumped Elements for RF and Microwave Circuits*. Artech House: Norwood, MA.

Bahl, I. J. and P. Bhartia, 1988. *Microwave Solid State Circuit Design*. 2nd edition. John Wiley & Sons: New York.

Bancroft, R., 2009. *Microstrip and Printed Antenna Design*. Institution of Engineering and Technology, SciTech Publishing: Delhi.

Behari, J., 2003. *Microwave Measurement Techniques and Applications*. Anamaya Publishers: New Delhi.

Benford, J., Swegle, J. A., and Schamiloglu, E., 2007. *High Power Microwaves*. CRC Press: Boca Raton, FL.

Benson, F. A. and Benson, T. M., 1991. *Fields, Waves and Transmission Lines*. Chapman & Hall: London, UK.

Bhartia, P. and Bahl, I. J., 1984. *Millimeter Wave Engineering and Applications*. John Wiley & Sons: New York.

Bhartia, P. and Pramanick, P., 1987. *E-Plane Integrated Circuits*. Artech House: Norwood, MA.

Bhat, B. and Koul, S. K., 1989. *Stripline-like Transmission Lines for Microwave Integrated Circuits*. New Age International Publishers: New Delhi.

Bhatt, N. D., 2004. *Elementary Engineering Drawing [Plane and Solid Geometry]*. Charotar Publishing House: Gujarat

Budimir, D., 1998. *Generalized Filter Design by Computer Optimization*. Artech House: Norwood, MA.

Caloz, C. and Itoh, T., 2005. *Electromagnetic Metamaterials: Transmission Line Theory and Microwave Applications*. John Wiley & Sons: Hoboken, NJ.

Carr, J. J., 2001. *Secrets of RF Circuit Design*. McGraw-Hill: New York.

Carson, R. S., 1982. *High-Frequency Amplifiers*. John Wiley & Sons: New York.

Chang, K., 2004. *RF and Microwave Wireless Systems*. John Wiley & Sons: New York.

Chaparro, L. F., 2014. *Signals and Systems Using MATLAB®*. 2nd edition. Elsevier: Amsterdam.

Chatterjee, R., 1992. *Microwave and Millimeter-Wave Semiconductor Devices*. East-West Press: New Delhi.

Chatterjee, R., 1996. *Antenna Theory and Practice*. 2nd edition. New Age International Publishers: New Delhi.

Chen, L. F., Ong, C. K., Neo, C. P., Varadan, V. V., and Varadan, V. K., 2004. *Microwave Electronics: Measurement and Materials Characterization*. John Wiley & Sons: New York.

Christodoulou, C. and Georgiopoulos, M., 2001. *Applications of Neural Networks in Electromagnetics*. Academic Press: Oxford, UK.

D'Ignazio, F. and Wold, A., 1984. *Artificial Intelligence (Franklin Watts Computer Library)*. The Watts Publishing Group: London, UK.

da Silva, E., 2001. *High Frequency and Microwave Engineering*. Butterworth-Heinemann: Oxford, UK.

Daintith, J. and J.O. Clark, 1999. *Universities Press Dictionary of Mathematics*. Universities Press: Hyderabad.

Das, A. and Das, S. K., 2000. *Microwave Engineering*. McGraw-Hill, New Delhi.

Davis, W. A., 1984. *Microwave Semiconductor Circuit Design*. Van Nostrand Reinhold: New York.

de los Santos, H. J., 2004. *Introduction to Microelectromechanical Microwave Systems*. Artech House: Norwood, MA.

Deb, G. K., Electromagnetic Interference and Electromagnetic Compatibility. Tata McGraw-Hill: New Delhi.

Devi, N., 2006. *Medicine in South India*. Eswar Press: Chennai.

Digit, 2005. *Fast Track to Wireless Networking*. Jasubhai Digital Media: Mumbai.

Dobrowolski, J. and W. Ostrowski, 1996. *Computer-Aided Analysis, Modeling, and Design of Microwave Networks: The Wave Approach*. Artech House: Norwood, MA.

Domach, M. M., 2004. *Introduction to Biomedical Engineering*. Pearson Education: London, UK.

Dorf, R. C., 2006. *The Electrical Engineering Handbook*, vol. 2. CRC Press: Boca Raton, FL.

Dube, D. C., 2011. *Microwave Devices and Applications*. Alpha Science International.

Edminister, J. A., 1979. *Theory and Problems of Electromagnetics*. McGraw-Hill: New York.

Edwards, T. C., 1992. *Foundations for Microstrip Circuit Design*. 2nd edition. John Wiley & Sons: New York.

Ellinger, F., 2008. *Radio Frequency Integrated Circuits and Technologies*. 2nd edition. Springer: New York.

Elliott, R. S., 1993. *An Introduction to Guided Waves and Microwave Circuits*. Prentice Hall: Upper Saddle River, NJ.

Enderle, J., Bronzino, J., and Blanchard, S. M., 2005. *Introduction to Biomedical Engineering*. Academic Press: Oxford, UK.

Engen, G. F., 1992. *Microwave Circuit Theory and Foundation of Microwave Metrology*. Peter Peregrinus: London, UK.

Everitt, W. L. and Anner, G. E., 1956. *Communication Engineering*. McGraw-Hill: New York.

Feynman, R. P., Gottlieb, M. A., and Leighton, R., 2013. *Feynman's Tips on Physics*. Pearson: New York.

Frey, J. and Bhasin, K. B., 1977. *Microwave Integrated Circuits*. Artech House: Norwood, MA.
Fusco, V. F., 1987. *Microwave Circuits: Analysis and Computer-Aided Design*. Pearson Education: London, UK.
Gandhi, O. P., 1995. *Microwave Engineering and Applications*. Pergamon Press: Oxford, UK.
Gardiol, F., 1994. *Microstrip Circuits*. John Wiley & Sons: New York.
Gardner, J. W., Varadan, V. K., and Awadelkarim, O. O., 2001. *Microsensors, MEMS, and Smart Devices*. John Wiley & Sons: New York.
Gibaldi, J., 1999. *MLA Handbook for Writers of Research Papers*. 5th edition. Modern Language Association of America: New York.
Gilmour, A. S., 1986. *Microwave Tubes*. Artech House: Norwood, MA.
Glover, I. A., Pennock, S., and Shepherd, P., 2006. *Microwave Devices Circuits and Subsystem*. John Wiley & Sons: New York.
Goldman, L., 2013. *The Biomedical Laser: Technology and Clinical Applications*. Springer: New York.
Gonzalez, G., 1997. *Microwave Transistor Amplifiers: Analysis and Design*. Prentice Hall: Upper Saddle River, NJ.
Gonzalez, R. C., Woods, R. E., and Eddins, S. L., 2004. *Digital Image Processing Using MATLAB®*. 2nd edition. Pearson Education: London, UK.
Great Britain War Office, 1955. *Royal Signals Handbook of Line Communication*. Standard Publishers Distributors: New Delhi.
Gunston, M. A. R., 1972. *Microwave Transmission-Line Impedance Data*. Van Nostrand Reinhold: New York.
Gupta, K. C., 1979. *Microwaves*. New Age International Publishers: New Delhi.
Gupta, K. C. and Singh, A., 1974. *Microwave Integrated Circuits*. John Wiley & Sons: New York.
Gupta, K. C., Garg, R., and Bahl, I. J., 1979. *Microstrip Lines and Slotlines*. Artech House: Norwood, MA.
Gupta, K. C., Garg, R., and Chadha, R., 1981. *Computer Aided Design of Microwave Circuits*. Artech House: Norwood, MA.
Gupta, S., 1977. *Microwave Engineering*. Khanna Publishers: Delhi.
Hall, P. S. and Jackson, D. R., 1993. *CAD of Printed Antennas and Array: Part 1. International Journal of Microwave and Millimeter-Wave Computer-Aided Engineering*, vol. 3, number 4. John Wiley & Sons: New York.
Harish, A. R. and Sachidananda, M., 2007. *Antennas and Wave Propagation*. Oxford, UK.
Harvey, A. F., 1963. *Microwave Engineering*. Academic Press: Oxford, UK.
Helszajn, J., 1978. *Passive and Active Microwave Circuits*. John Wiley & Sons: New York.
Hoffmann, R. K., 1987. *Handbook of Microwave Integrated Circuits*. Artech House: Norwood, MA.
Hong, J.-S. and Lancaster, M. J., 2001. *Microstrip Filters for RF/Microwave Applications*. John Wiley & Sons: New York.
Howe, H. H., 1985. *Stripline Circuit Design*. Microwave Associates: Baltimore, MD.
Hsu, T.-R., 2002. *MEMS and Microsystems: Design and Manufacture*. McGraw-Hill: New York.
Hudson, D. L. and Cohen, M. E., 2000. *Neural Networks and Artificial Intelligence for Biomedical Engineering*. John Wiley & Sons: New York.
IEEE, 2004. Protecting the homeland: The many facets of homeland security. *IEEE Engineering in Medicine and Biology Magazine*, vol. 23, number 1.

IEEE, 2007. *IEEE Transactions on Microwave Theory and Techniques*, vol. 55, number 6. IEEE: Piscataway, NJ.

IEEE, 2008. MRI robotics: Going beyond traditional diagnostics with revolutionary systems. *IEEE Engineering in Medicine and Biology Magazine*, vol. 27, number 3, 108 pages.

IEEE, 2008. Uninteroperability: Problems perpetuated by functional disconnection. *IEEE Engineering in Medicine and Biology Magazine*, vol. 27, number 6.

IEEE, 2015. Waves all around: Networking in the underground. *IEEE Antennas & Propagation Magazine*, vol. 57, number 4.

IEEE, 2018. Exploring the potential of OAM antennas. *IEEE Antennas & Propagation Magazine*, vol. 60, number 2.

IEEE, 2018. The breadth of MTT: Multiple methods, common aims. *IEEE Microwave for the Microwave & Wireless Engineer Magazine*, vol. 19, number 4.

IETE, 1988. *IETE Technical Review*, vol. 5, number 1. IETE: New Delhi.

IETE, 1998. *IETE Technical Review*, vol. 15, number 4. IETE: New Delhi.

Ishii, T. K., 1989. *Microwave Engineering*. Harcourt Brace Jovanovich: San Diego, CA.

Ishii, T. K., 1995. *Handbook of Microwave Technology*, vol. 2. Academic Press: Oxford, UK.

Itoh, T., 1989. *Numerical Techniques for Microwave and Millimeter-Wave Passive Structures*. John Wiley & Sons: New York.

Jacobson, B. and Webster, J. G., 1977. *Medicine and Clinical Engineering*. Prentice Hall: Upper Saddle River, NJ.

Jain, P. K. and Kaur, G., 1994. *Networks, Filters and Transmission Lines*. McGraw-Hill: New Delhi.

James, J. R., Hall, P. S., and Wood, C., 1986. *Microstrip Antenna: Theory and Design*. Peter Peregrinus: London, UK.

Jog, N. K., 2006. *Electronics in Medicine and Biomedical Instrumentation*. Prentice-Hall: New Delhi.

Jordan, E. C. and Balmain, K. G., 1968. *Electromagnetic Waves and Radiating Systems*. Prentice Hall: Upper Saddle River, NJ.

Kalechman, M., 2018. *Practical MATLAB Applications for Engineers*. CRC Press: Boca Raton, FL.

Kennedy, G., 1985. *Electronic Communication Systems*. 3rd edition. McGraw-Hill: New Delhi.

King, R. W. P. and Harrison, C. W., 1969. *Antennas and Waves: A Modern Approach*. MIT Press: Cambridge, MA.

Kinsler, L. E. and Frey, A. R., 1950. *Fundamentals of Acoustics*. John Wiley & Sons: New York.

Kneppo, I. and Fabian, J., 2012. *Microwave Integrated Circuits*. Springer: New York.

Koryu Ishii, T., 1995. *Handbook of Microwave Technology: Components and Devices*, vol. 2. Academic Press: Oxford, UK.

Koul, S. K. and Bhat, B., 1991. *Microwave and Millimeter Wave Phase Shifters: Dielectric and Ferrite Phase Shifters*, vol. 1. Artech House: Norwood, MA.

Koul, S. K. and Bhat, B., 1997. *Computer-Aided Design of Millimeter Wave Fin lines: Analysis and Design Software for Windows*. New Age Publishers: New Delhi.

Kraus, J. D., Marhefka, R. J., and Khan, A. S., 2006. *Antennas and Wave Propagation*. 4th edition. McGraw-Hill: New Delhi.

Laverghetta, T. S., 1987. *Solid State Microwave Devices*. Artech House: Norwood, MA.

Laverghetta, T. S., 1998. *Microwaves and Wireless Simplified*. Artech House: Norwood, MA.

Lee, C. A. and Dalman, G. C., 1994. *Microwave Devices, Circuits and Their Interaction.* John Wiley & Sons: New York.

Lee, T. H., 2004. *Planar Microwave Engineering: A Practical Guide to Theory, Measurement, and Circuits.* Cambridge University Press: Cambridge, MA.

Lehpamer, H., 2010. *Microwave Transmission Networks: Planning, Design, and Deployment.* 2nd edition. McGraw-Hill: New Delhi.

Liao, S. Y., 1987. *Microwave Circuit Analysis and Amplifier Design.* Prentice Hall: Upper Saddle River, NJ.

Liao, S. Y., 1990. *Microwave Devices and Circuits.* Prentice Hall: Upper Saddle River, NJ.

Lioubtchenko, D., Tretyakov, S., and Dudorov, S., 2003. *Millimeter-Wave Waveguides.* Kluwer Academic Publishers: New York.

Liu, C., 2012. *Foundations of MEMS.* Pearson Education: London, UK.

Ludwig, R. and Bogdanov, G., 2000. *RF Circuit Design: Theory & Applications.* 2nd edition. Pearson Education: London, UK.

Ludwig, R. and Bretchko, P., 2000. *RF Circuit Design: Theory and Applications.* Pearson Education: London, UK.

Maas, S. A., 1998. *The RF and Microwave Circuit Design Cookbook.* Artech House: Norwood, MA.

Madou, M. J., 1997. *Fundamentals of Microfabrication.* CRC Press: Boca Raton, FL.

Magnusson, P. C., 1965. *Transmission Lines and Wave Propagation.* Allyn and Bacon: Boston, MA.

Makimoto, M. and Yamashita, S., 2013. *Microwave Resonators and Filters for Wireless Communication: Theory, Design, and Applications.* Springer: New York.

Maloratsky, L. G., 2003. *Passive RF and Microwave Integrated Circuits.* Harcourt India Pvt Ltd: New Delhi.

Matthaei, G. L., Young, L., and Jones, E. M. T., 1980. *Microwave Filters, Impedance-Matching Networks, and Coupling Structures.* Artech House: Norwood, MA.

McKenzie, B. C., 1997. *Medicine and the Internet: Introducing Online Resources and Terminology.* Oxford University Press: Oxford, UK.

Medley, M. W., 1993. *Microwave and RF Circuits: Analysis, Synthesis, and Design.* Artech House: Norwood, MA.

Milligan, T. A., 2005. *Modern Antenna Design.* 2nd edition. John Wiley & Sons: New York.

Millman, J. and Taub, H., 2001. *Pulse, Digital and Switching Waveforms.* McGraw-Hill: New Delhi.

Misra, D. K., 2004. *Radio-Frequency and Microwave Communication Circuits: Analysis and Design.* John Wiley & Sons: Hoboken, NJ.

Mittra, R., 1973. *Computer Techniques for Electromagnetics.* Pergamon Press: Oxford, UK.

Nair, B. S., 2008. *Microwave Engineering: Theory, Analyses and Applications.* Sanguine Technical Publishers: Bangalore.

Nelson, C., 2018. *High-Frequency and Microwave Circuit Design.* 2nd edition. CRC Press: Boca Raton, FL.

Nguyen, C., 2003. *Analysis Methods for RF, Microwave and Millimeter-Wave Planar Transmission Line Structures.* John Wiley & Sons: New York.

O'Connor, H. N. and Kaisha, S. D. K., 1981. *Microwave Miracles, Variable-Power, from Sanyo.* B. R. Publishing Corporation: New Delhi.

Ogata, K., 2010. *Modern Control Engineering.* Prentice Hall: Upper Saddle River, NJ.

Pennock, S. R. and Shepherd, P. R., 1988. *Microwave Engineering with Wireless Applications.* MacMillan Press: London, UK.

Philips Natuurkundig Laboratorium, 1971. *Philips Technical Review*, vol. 32. Philips Research Laboratory: USA.

Prasad Kodali, V., 2001. *Engineering Electromagnetic Compatibility: Principles, Measurements, Technologies, and Computer Models*. Wiley India Pvt Ltd: New Delhi.

Pratap, R., 2006. *Getting Started with MATLAB 7: A Quick Introduction for Scientists and Engineers*. Oxford University Press: Oxford, UK.

Prince, J. L. and Links, J. M., 2008. *Medical Imaging Signals and Systems*. Pearson Education: London, UK.

Radmanesh, M. M., 2001. *Radio Frequency and Microwave Electronics Illustrated*. Prentice Hall: Upper Saddle River, NJ.

Raghavan, S., 1999. Characterization of discontinuities in shielded coplanar waveguide. *Doctor of Philosophy*, Indian Institute of Technology, Delhi.

Raja Rao, C. and Guha, S. K., 2001. *Principles of Medical Electronics and Biomedical Instrumentation*. Universities Press: Hyderabad.

Rajendran, V., Hillebrands, B., Prabu, P., and Geckeler, K. E., 2011. *Biomedical Applications of Nanostructured Materials*. Macmillan Publishers India Limited: Noida.

Ramakrishna, S., Ramalingam, M., and Sampath Kumar, T. S., 2016. *Biomaterials: A Nano Approach*. CRC Press: Boca Raton, FL.

Ratner, B. D., Hoffman, A. S., and Schoen, F. J., 2012. *Biomaterials Science: An Introduction to Materials in Medicine*. 3rd edition. Academic Press: Oxford, UK.

Rauscher, C., Janssen, V., and Minihold, R., 2007. *Fundamentals of Spectrum Analysis*. Rohde & Schwarz: Munich.

Reddy, D. C., 2005. *Biomedical Signal Processing: Principles and Techniques*. McGraw-Hill: New Delhi.

Richards, D., Clarke, C. E., and Clark, T., *The Human Brain and Its Disorders*. Oxford University Press: Oxford, UK.

Rizzi, P. A., 1988. *Microwave Engineering: Passive Circuits*. Prentice Hall: Upper Saddle River, NJ.

Rohde, U. L. and Newkirk, D. P., 2000. *RF/Microwave Circuit Design for Wireless Applications*. John Wiley & Sons: New York.

Roy, S. K. and Mitra, M., 2003. *Microwave Semiconductor Devices*. PHI Learning Pvt. Ltd.: New Delhi.

Ryder, J. D., 1961. *Networks, Lines and Fields*. 2nd edition. Pearson Education India: Chennai.

Saad, T. S., 1971. *Microwave Engineers Handbook*, vol. 2. Artech House: Norwood, MA.

Sadiku, M. N. O., 2009. *Principles of Electromagnetics*. Oxford University Press: Oxford, UK.

Saliterman, S. S., 2006. *Fundamentals of BioMEMS and Medical Microdevices*. SPIE Press: Washington, DC.

Sanei, S. and Chambers, J. A., 2007. *EEG Signal Processing*. John Wiley & Sons: New York.

Sazonov, D. M., 1982. *Microwave Circuits and Antennas*. Mir Publishers: Moscow.

Silver, S., 1984. *Microwave Antenna Theory and Design*. Peter Peregrinus Ltd: London, UK.

Simons, R. N., 2004. *Coplanar Waveguide Circuits, Components, and Systems*. John Wiley & Sons: New York.

Sivanandam, S. N., Sumathi, S., and Deepa, S. N., 2006. *Introduction to Neural Networks Using Matlab 6.0*. McGraw-Hill: New York.

Sörnmo, L. and Laguna, P., 2005. *Bioelectrical Signal Processing in Cardiac and Neurological Applications*. Academic Press: Oxford, UK.

Soundararajan, V., 2002. *Antenna Theory and Wave Propagation*. SciTech Publications: New Delhi.

Spohr, M. H., 1983. *The Physician's Guide to Desktop Computers*. Reston Publishing Company: Reston, VA.

Steer, M. B., 2010. *Microwave and RF Design: A Systems Approach*. SciTech Publications: New Delhi.

Stratton, J. A., 2007. *Electromagnetic Theory*. John Wiley & Sons: New York.

Subbarao, V., 2001. *Numerical Methods in Electromagnetic Fields*. Alpha Science International.

Terman, F. E., 1984. *Electronic and Radio Engineering*. McGraw-Hill: New York.

Uher, J., Bornemann, J., and Rosenberg, U., 1993. *Waveguide Components for Antenna Feed Systems: Theory and CAD*. Artech House: Norwood, MA.

Vander Vorst, A., Rosen, A., and Kotsuka, Y., 2006. *RF/Microwave Interaction with Biological Tissues*. John Wiley & Sons: New York.

Vasuki, S., Helena Margaret, D., and Rajeswari, R., 2015. *Microwave Engineering*. McGraw-Hill: New Delhi.

Vendelin, G. D., Pavio, A. M., and Rohde, U. L., 2005. *Microwave Circuit Design Using Linear and Nonlinear Techniques*. 2nd edition. John Wiley & Sons: New York.

Viswanathan, T., 1992. *Telecommunication Switching Systems and Networks*. PHI Learning: New Delhi.

Vizmuller, P., 1995. *RF Design Guide: Systems, Circuits, and Equations*. Artech House: Norwood, MA.

Wadell, B. C., 1991. *Transmission Line Design Handbook*. Artech House: Norwood, MA.

Walker, J. L. B., 2006. *Classic Works in RF Engineering: Combiners, Couplers, Transformers, and Magnetic Materials*. Artech House: Norwood, MA.

Weisman, C. J., 2002. *The Essential Guide to RF and Wireless*. Pearson Education: London, UK.

Wolff, I., 2006. *Coplanar Microwave Integrated Circuits*. John Wiley & Sons: Hoboken, NJ.

Wong, K.-L., 2004. *Compact and Broadband Microstrip Antennas*. John Wiley & Sons: New York.

Young, L., 1972. *Microwave Filters Using Parallel Coupled Lines*. Artech House: Norwood, MA.

Bibliography

Prof. S. Raghavan gratefully acknowledges the author(s) of the following books which are referred.

Inder Bahl and Prakash Barthia (2003). *Solid State Circuit Design*, Wiley, New York.

Christos Christodoulou and Michael Georgiopoulos (2000). *Applications of Neural Networks in Electromagnetics*, Artech House Inc., Norwood, MA.

Yoshihiro Konishi (2009). *Microwave Integrated Circuits*, Marcel Dekker Inc., New York.

Leo G. Maloratsky (2003). *Passive RF and Microwave Integrated Circuits*, Elsevier, New Delhi.

P. I. Somlo and John D. Hunter (1985). *Microwave Impedance Measurement*, Peter Peregrinus Limited, London.

Index

M

N

O

P

T

U

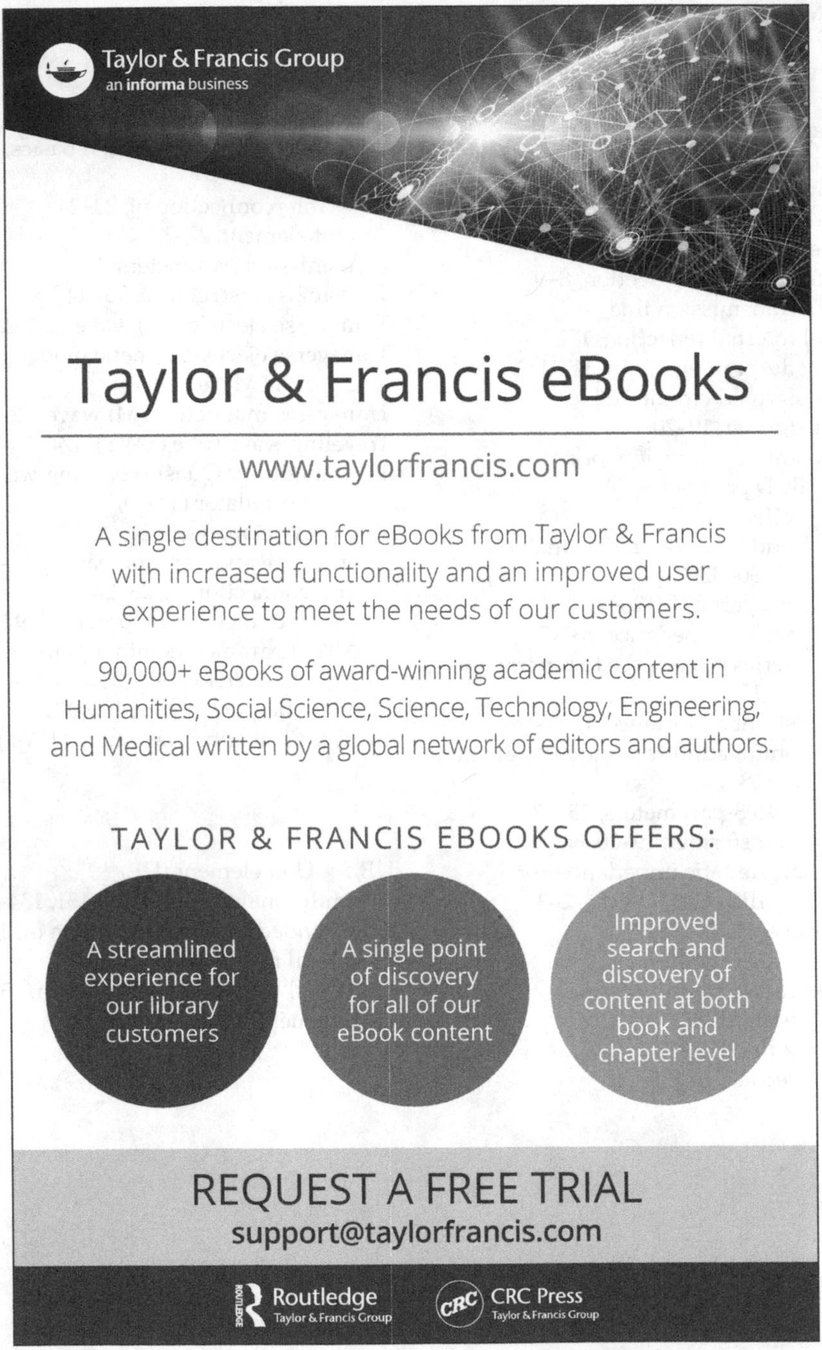
Taylor & Francis Group
an informa business
Taylor & Francis eBooks
www.taylorfrancis.com
A single destination for eBooks from Taylor & Francis with increased functionality and an improved user experience to meet the needs of our customers.
90,000+ eBooks of award-winning academic content in Humanities, Social Science, Science, Technology, Engineering, and Medical written by a global network of editors and authors.
TAYLOR & FRANCIS EBOOKS OFFERS:
A streamlined experience for our library customers
A single point of discovery for all of our eBook content
Improved search and discovery of content at both book and chapter level
REQUEST A FREE TRIAL
support@taylorfrancis.com
Routledge
Taylor & Francis Group
CRC Press
Taylor & Francis Group

Printed in the United States
by Baker & Taylor Publisher Services

Printed in the United States
by Baker & Taylor Publisher Services